OXFORD NEUROSCIENCE SERIES

Series Editors

OXFORD NEUROSCIENCE SERIES

1 *Calcium ions in nerve cell function*
Platon G. Kostyuk

Calcium Ions in Nerve Cell Function

PLATON G. KOSTYUK

A. A. Bogomoletz Institute of Physiology
Kiev, Ukraine

Oxford New York Tokyo
OXFORD UNIVERSITY PRESS
1992

Oxford University Press, Walton Street, Oxford OX2 6DP
Oxford New York Toronto
Delhi Bombay Calcutta Madras Karachi
Petaling Jaya Singapore Hong Kong Tokyo
Nairobi Dar es Salaam Cape Town
Melbourne Auckland
and associated companies in
Berlin Ibadan

Oxford is a trade mark of Oxford University Press

Published in the United States
by Oxford University Press, New York

A catalogue record for this book is available from the British Library

Library of Congress Cataloging in Publication Data
Kostyuk, P. G. (Platon Grigor'evich)
Calcium ions in nerve cell function/Platon G. Kostyuk.
(Oxford neuroscience series: 1)
Includes bibliographical references and index.
1. Neurons. 2. Calcium—Physiological effect. 3. Calcium channels. I. Title. II. Series.
[DNLM: 1. Calcium—pharmacology. 2. Calcium Channels—physiology. 3. Neurons—physiology. WL 102.5 K86c]
QP363.K627 1992 591.1'88—dc20 92-24093
ISBN 0-19-854672-6

Typeset by Joshua Associates Ltd, Oxford
Printed in Great Britain by
St. Edmundsbury Press, Bury St Edmunds, Suffolk

Preface

The idea that monovalent and divalent ions play a different role in the maintenance of the basic functional properties of living cells, and especially of their capability to respond by an active reaction (excitation) to an external stimulus, has a century-old history. Due to the pioneering observations of Ringer (1883) and Loeb (1906) it became clear at the end of the nineteenth century that a definite relationship between both types of ions is necessary for maintaining normal cellular activity (for instance, for cellular contraction). Loeb stated quite definitely that as soon as the tissue becomes free from calcium ions, sodium ions immediately lose their ability to induce contraction. After addition of excessive calcium salts to sodium chloride solution, rhythmic muscle contractions stop completely. These findings indicate that contractions can be evoked only at a given concentration of sodium ions: however, if the sodium to calcium ratio becomes too high, contractions become impossible; they are also impossible when this ratio is too low. At the same time it became clear that magnesium ions also act in an inhibitory way, like calcium ions, and that the antagonism between mono- and divalent ions also manifests itself in the functions of other excitable tissues.

The discovery that electric current is a universal stimulator of excitable tissues greatly enlarged the experimental possibilities for precise measurements of the functional properties of these tissues. The antagonistic action of an increase in the concentration of divalent cations on electrical stimulation was established by Loeb and many other investigators; the necessity of a definite relation between the concentrations of mono- and divalent cations has even been considered by some Russian biophysicists as the 'main law of excitation' (Lazarev 1923). The fundamental reasons for such antagonism, as well as for the origin of bio-electric phenomena themselves, were for a long time related to different mobilities of mono- and divalent cations (see, for example, Chagovetz 1903–1906). However, the insufficiency of simple diffusional approaches became more and more obvious. Finally, the introduction of the concept of a semipermeable surface membrane in every cell (Bernstein 1912) created a real basis for the understanding of ionic mechanisms of cellular excitability which have been so successfully investigated ever since.

During the last few decades we have witnessed great progress in the investigation of the physico-chemical basis of living events, due to the wide introduction into physiology of new methods of structural, electrophysiological, and biochemical analysis. The mechanism of initiation and propagation of the nerve impulse—the main active response of the nerve cell—have become clear in all details; the nature of the chemosensitive properties of the cell membrane and the mode of intercellular transmission of signals have been well established; the main principles of coupling between the excitatory and effector processes (contraction, secretion) have also been determined. In all these phenomena we always meet with the problem of ionic antagonism discovered by our great predecessors, although on a completely new level.

A special role of divalent cations in the functioning of excitable cells reveals itself in a variety of ways; it almost seems that there is no a single link in cellular activity in which these ions do not participate. Recent advances in bio-organic chemistry have opened a way for the understanding of this universal role of calcium ions—by showing their highly specific binding to most important macromolecules, followed by corresponding changes in cellular functions controlled by them.

The relative roles of mono- and divalent cations in the mechanisms of cellular excitability have been studied in the Department of General Physiology, Bogomoletz Institute of Physiology, Ukrainian Academy of Sciences (Kiev) for many years, using the most advanced technical approaches, and a lot of experimental data have been obtained, some of which is of key importance. The present book summarizes these findings and compares them with those obtained in other laboratories. Therefore the reader can obtain an up-to-date picture of the contemporary status of each question discussed.

Kiev, Ukraine P.G.K.
September 1991

Contents

1. Action of calcium ions on the surface of nerve cells

History

Modern ideas about the mechanisms of interaction between calcium ions and the external surface of excitable cell membranes originate from the studies of Frankenhaeuser (1957) on frog nerve fibres and Frankenhaeuser and Hodgkin (1957) on squid giant axons. It was established in these experiments that an increase in the extracellular Ca^{2+} concentration greatly increases the threshold of action potential generation without changing the resting potential value. A similar alteration has been found in other excitable structures, for instance in cardiac muscle fibres (Weidman 1955).

As has since been established, the reason for such a change is the existence of fixed charged groups on the surface of cell membranes. The charges of these groups do not contribute by themselves to the transmembrane potential difference, which is based on the diffusion of ions. However, in combination with diffusion potentials the surface charges can change the transmembrane electric field.

If fixed charges on the membrane surface were always constant, their presence could be neglected when studying membrane processes. However, it is clear that changes in the characteristics of the solutions bathing the membrane surface (ionic composition, ionic strength, pH) will alter the value of the surface potential created by these charges and will correspondingly complicate the interpretation of functional alterations in membrane mechanisms produced by such changes.

The quantitative description of surface potentials and their changes is based on the idea about a double-electric layer elaborated by Gouy (1910) and Chapman (1913) and then essentially modified by Stern (1924). According to these ideas, introduction of certain ions into the solution bathing the charged membrane surface leads to a change in its potential due to the formation of a screening layer of counterions, on the one hand, and to neutralization of charges of corresponding groups, on the other.

The screening effect can be described by the Gouy–Chapman equation:

$$\sigma = \sqrt{\left[\frac{\varepsilon RT}{2\pi} \sum C_i\{\exp(-z_i \varphi F/RT) - 1\}\right]} \tag{1.1}$$

which relates the value of surface potential φ to the density of fixed charges σ (the latter represents the charges of anionic groups in biological membranes, as will be shown below, and is consequently negative), concentration of counterions in the solution C_i, and their valency z_i.

Changes in the membrane surface charge due to the screening effect of the double diffusion layer will inevitably cause the changes inside the membrane field and, consequently, will lead to the changes in the mechanisms of potential-dependence, which provide excitation of the membrane. This mechanism was used to explain the effect of pH change of the extracellular solution and of increased external content of divalent cations on the functional characteristics of the neuronal membrane (Gilbert and Ehrenstein 1969; McLaughin *et al.* 1971; Mozhayeva and Naumov 1972*a*, *b*).

However, explanation of the changes in the membrane surface potential only by the screening effect of counterions cannot be sufficient; it is clear that binding of counterions by surface groups also takes place. Such binding will reduce the density of surface charges and, correspondingly, the surface potential. At the same time, with the decrease of surface charges, electrostatic attraction of counterions from the solution will be weakened and their accumulation near the membrane surface will be decreased. A quantitative description of ion binding can be given using the Langmuir isotherm:

$$\sigma = \frac{\sigma_0}{1 + [M_i]/K_d^i} \tag{1.2}$$

where K_d^i is the apparent dissociation constant for the complexes formed by a given ion (in most papers the inverse association constant expressed in M^{-1} is used, but the use of K_d is more convenient for comparison with other physiological data), $[M_i]$ is its near-membrane concentration, σ_0 is the charge density with completely free groups, and σ is the density after binding. If the solution contains several sorts of binding ions, the participation of each ion in the neutralization of surface charges will be proportional to both their concentrations and binding constants. Near-membrane concentration of the ion differs from its volumetric concentration, C, by a value:

$$[M_i] = C_i \exp(-z_i \varphi F/RT). \tag{1.3}$$

Therefore, the equation (1.2) can be written as follows:

$$\sigma = \frac{\sigma_0}{\{1 + C_i/K_d^i\}\{\exp(-z_i \varphi F/RT)\}} \tag{1.4}$$

If both screening and ion binding take place, both equations have to be combined in one equation:

$$\frac{\sigma_0}{\{1 + C_i/K_d^i\}\{\exp(-z_i \varphi F/RT)\}} = \sqrt{\left[\frac{\varepsilon RT}{2\pi} \sum C_i\{\exp(-z\varphi F/RT) - 1\}\right]} \tag{1.5}$$

which is used when analysing experimental data. If ions present in the solution exert their effect only by screening of surface charges, only their valency and concentration would be of importance in this case; if binding takes place, the effect should also depend on chemical characteristics of the ions, which determine the effectiveness of their binding to fixed groups. In studies on the Ranvier nodes of frog nerve fibres, equal concentrations of different divalent cations have been shown to have different effects on potential-dependent characteristics of the surface membrane, i.e. they must associate themselves with fixed groups to a different extent. According to the efficiency of binding, these ions can be placed in the following sequence (Hille 1968; Hille *et al.* 1975):

$$Zn^{2+} > Ni^{2+} > Mn^{2+} > Co^{2+} \cong Ca^{2+} > Mg^{2+} > Sr^{2+} = Ba^{2+}.$$

For the membrane of lobster axon, this series turned out to be somewhat different (Blaustein and Goldman 1968):

$$La^{3+} > Ni^{2+} > Zn^{2+} \cong Co^{2+} > Cd^{2+} \cong Ba^{2+} > Ca^{2+}.$$

Theoretical consideration of a possible process of ion binding to fixed groups, and of the changes in the density of surface charges resulting from such a binding, reveals difficulties (see Kostyuk *et al.* 1982*b*). Thus, such binding may have a number of forms; the extreme cases are the formation of charged complexes of ions M^{z+} and groups X^- of 1:1 type ($M^{z+} + X = MX^{(z-1)+}$) and neutral complexes whose stoichiometry corresponds to the valency of the ion ($M^{z+} + zX = MX_z$). In the latter case, such complexes will be neutral and, as they are formed, the density of surface charges will tend to zero. In the former case, not only neutralization of charges but also a recharge of the membrane surface can be expected; Intermediate cases are also possible. The calculations will be different depending on the adoption of one or the other suggestion. Based on the data obtained from the studies of artificial phospholipid membranes containing fixed charges, there is an impression that polyvalent ions form complexes of mainly the MX_z type (McLaughlin *et al.* 1971). However, the same authors observed (using the microelectrode technique) the possibility of recharging the surface of phosphatidyl serine vesicles upon introduction of divalent cations into the medium at a concentration of about 0.1 M (McLaughlin *et al.* 1981). However, such high concentrations of divalent ions may produce side-effects (aggregation of vesicles and others).

Therefore, the second suggestion seems to be more justified and we may assume that formation of complexes of the MX_z type prevails (see below). We may also assume for simplicity that only protons and di- and also polyvalent cations bind to surface groups. This simplification is based on the fact that the effectiveness of binding for monovalent cations with different anionic groups is 1–2 orders of magnitude less than that for di- and polyvalent cations (see Martell and Smith 1977). If the concentration of both protons and divalent cations in the medium is changed simultaneously, it is likely that initially HX complexes are formed and then the remaining free X^- groups react with divalent cations and form MX_z complexes.

Modern methods for the quantitative study of electrical processes on cell membranes: intracellular perfusion of isolated cells

In our studies on ion interactions with nerve cell membranes isolated intracellularly, perfused cells were mostly used. Such experimental conditions allowed us to conduct very accurate measurements of ionic currents in the cell membrane and to strictly control the environment at its surface. We shall come back several times to the technique of intracellular perfusion, and so will describe it in more detail.

The main principle of the method (Kostyuk *et al.* 1975) is as follows. An isolated cell is placed into a pore (whose diameter is less than that of the cell) made in the tip of a plastic pipette. The cell is taken up into the pore by a flow of the solution due to the difference of hydrostatic pressures created between the chamber in which the cell is placed and the pipette. Then a part of the cell membrane facing the future 'intracellular' solution inside the pipette is destroyed by either a fast jump of hydrostatic pressure or mechanically.

Electrical processes on the membrane whose external and internal parts are in contact with extra- and intracellular solutions, respectively, can be recorded with the help of a measuring set-up connected to the solutions in the pipette and in the chamber. The ruptured membrane sticks to the pore walls, thus preventing leakage of current between the compartments and spontaneous closure of the hole. The elimination of external diffusion barriers due to cell isolation allows a fast and reliable control of the medium composition at the external membrane surface. Through the hole in the membrane one can quickly change the medium at its inner surface; it is best to define such a change as cell dialysis.

The presence of a large series resistance created by the pore is the main shortcoming of the method in its simplest form. This resistance results in prolongation of the time of recharge of the membrane capacitance and possible error during potential control under voltage-clamp conditions. This drawback can be overcome to a considerable extent by modification of the method.

In this case one pipette is connected with the perfusing system containing the intracellular solution; it can be easily supplied to any part of the cell by a micromanipulator. After fixation of the cell in the pore of the pipette, another similar pipette is brought in contact with the cell from the opposite side and a second opening is made in the cell membrane. Now a flow of the intracellular solution can be created through the cell by varying the hydrostatic pressure in both pipettes, i.e. a true cell perfusion can be achieved. In this modification the membrane potential can be measured by the electronic circuit connected to one pipette and the feedback current can be passed through the other one; in this way a considerable part of the series resistance can be removed and more precise electrical characteristics of the membrane can be obtained. The perfusion method is especially important for the investigation of large cells because it speeds up considerably the change of the intracellular medium; the large membrane capacitance of such cells makes them more sensitive to the effects of series resistance. Figure 1.1 presents schematically both approaches and the ways of connection of the electronic circuits.

The problem of structural changes in isolated neurons during cell dialysis

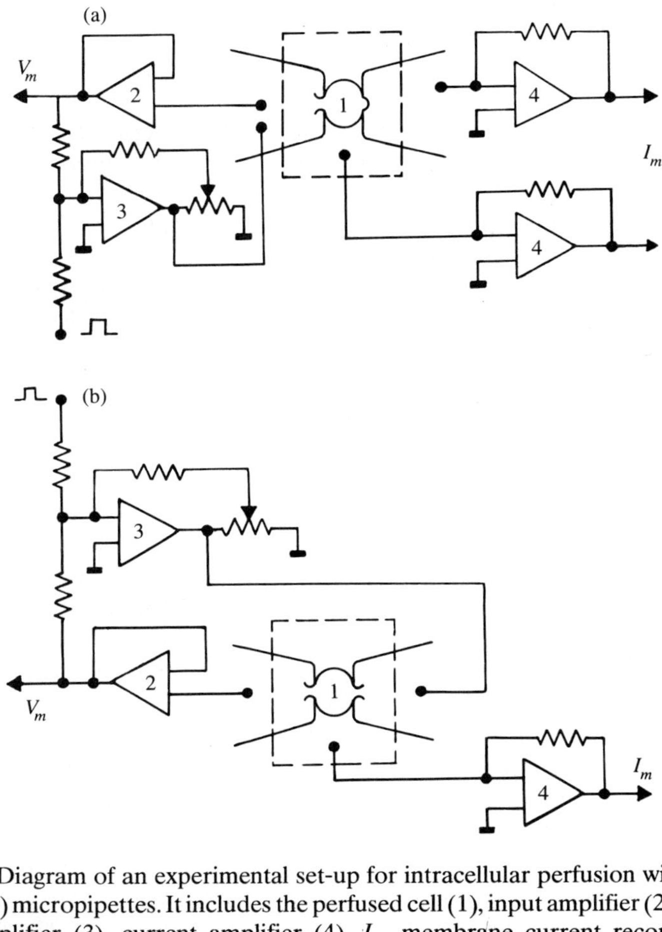

Fig. 1.1. Diagram of an experimental set-up for intracellular perfusion with one (a) and two (b) micropipettes. It includes the perfused cell (1), input amplifier (2), voltage-clamp amplifier (3), current amplifier (4). I_m, membrane current recording; V_m, membrane potential recording.

or perfusion is of major importance. To investigate them, electron microscopic observations were made on snail neurons fixed immediately after the beginning of perfusion. It has been shown that the main part of the cytoplasm as well as intracellular organelles are preserved in the cell. In this respect the situation here is different from the well-known perfusion of squid axons where the axoplasm is almost completely extruded and only a thin layer remains at the internal surface of the membrane. In perfused snail neurons the cytoplasmic matrix becomes highly vacuolized, probably due to distension of the endoplasmic reticulum and mitochondria. Such vacuolization may be an important factor facilitating fast exchange of intracellular soluble substances (Pogorelaya *et al.* 1980). The importance of this feature of cell perfusion will be considered in Chapter 6.

The perfusion of isolated cells can also be performed with glass micropipettes; the cell can be located either inside the pipette in its tip (Takahashi and Yoshii 1978) or outside (Lee *et al.* 1978). As in the previous case, a patch of the membrane at the opening of the pipette is destroyed and the opening is used for the replacement of the intracellular content. The presence of fixed charges on the glass may prevent the adhesion of the cell membrane to it; special pretreatment of glass (for instance with protamine) can be used to neutralize the charges.

The intracellular perfusion methods are described more extensively in several reviews (Kostyuk 1981, 1982; Kostyuk *et al.* 1981*a*) and also in the volume *Intracellular perfusion of excitable cells* (Kostyuk and Krishtal 1984).

As was already mentioned, full use of the advantages offered by intracellular perfusion can be achieved only on isolated neurons free of any barriers for the application of external substances. Such isolation meets certain difficulties which should be mentioned briefly. Amongst the most convenient cells for isolation are the neurons from mollusc ganglia. They are large in size and simple in shape, being free from dendrites. The procedure is based on a combination of enzymatic treatment, which softens the connective tissue around the neurons, and mechanical isolation of the cells. Different proteolytic enzymes can be used: trypsin, hyaluronidase, pronase (the latter, according to several authors, is the most effective). The duration of incubation with the enzyme depends on the density of the connective tissue; in all cases it should be minimal to prevent damage of neuronal membranes. This is especially important for the determination of surface groups of the membrane, which can be destroyed by the action of the enzymes; for this purpose complete elimination of the enzymatic step is recommended.

During mechanical separation of the neurons their axons are torn off from the somata; the opening in the membrane usually seals itself and the isolated cells retain their excitability for a long time if maintained in a suitable incubation medium. Pre-incubation in a fridge for 2–4 h helps to keep neurons alive. An isolated snail neuron fixed between two plastic perfusion pipettes is shown in Fig. 1.2.

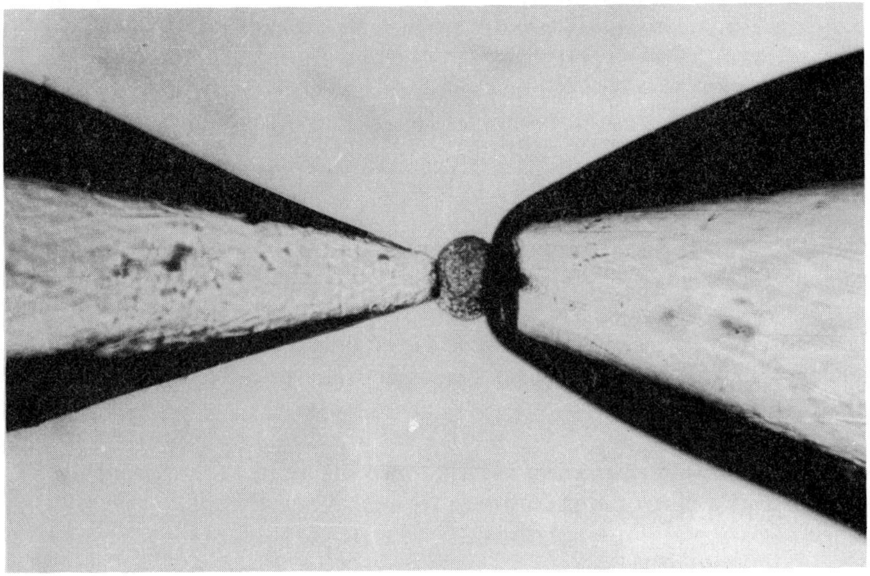

Fig. 1.2. Microphotograph of an isolated snail neuron fixed between two perfusing micropipettes. Cell diameter 90 μm.

Similar isolation techniques proved to be applicable to spinal ganglia of vertebrates, the neurons of which are also devoid of dendrites but are much smaller in size. Such an isolation technique is especially convenient for new-born rats, in which ganglia can be treated with pronase. In adult animals the connective tissue is considerably thicker, therefore damage to the cell membranes during isolation is more likely. This technique has also been used for the successful isolation of neurons from trigeminal and other sensory ganglia of brain nerves.

The isolation of cells from different divisions of the central nervous system is naturally of great interest. In this respect considerable success has been gained recently using brain slices. Such slices are pre-incubated for a while in normal saline solution at constant oxygenation and at a temperature of 37 °C and then are placed in a solution containing proteolytic enzymes (collagenase and trypsin) at concentrations of about 1 mg ml^{-1}. Keeping the tissue in such a solution (upon continuous oxygenation) makes the connective tissue soft, and in several dozens of seconds single neurons from the corresponding brain structure can be removed from the slice by a micropipette. The cells are gathered in saline solution containing 20 per cent bovine serum and can be used in experiments for up to 12 h after isolation. It is natural that a considerable proportion of the cell's processes are torn off or retracted, although parts of them are still preserved (see details in Kaneda *et al.* 1988).

The isolation of single heart cells has also been successfully achieved; in this case cardiac tissue is usually treated by collagenase at low calcium concentrations (Powell and Twist 1976; Clark *et al.* 1978; Rajs *et al.* 1978; Undrovinas 1984). The cells thus isolated are approximately cylindrical and are likely to be the result of complete dissociation of the myocardium into single cardiomyocytes.

It should be noted that there are convenient sources of single excitable cells which hardly require any preliminary isolation. These are cultures of neuroblastoma cells (from a malignant tumour of mouse sympathetic ganglia), phaeochromocytoma cells (from a malignant tumour of the adrenals), etc. Such cells have a simple shape and can be isolated easily from the bulk of culture. They preserve electrical excitability and chemoreceptive properties which are similar to those of normal excitable cells in many respects.

Experimental determination of calcium binding to the external surface of the neuronal membrane from shifts of the current–voltage characteristics of ionic currents

One of the widely used methods of indirect determination of the binding of divalent cations to the external surface of the cellular membrane is based on the measurement of the shifts that occur in the current–voltage curves for transmembrane ion currents. In our experiments isolated perfused neurons from snail (Kostyuk *et al.* 1982*a*, *b*) or rat (Kostyuk and Mironov 1982) have been used for this purpose. Perfusion washes out potassium ions from the cell and thus considerably decreases the outward currents. Therefore, in such cells the inward currents carried by sodium or calcium ions can be measured in almost pure form. For separate measurement of the sodium current, fluoride ions, which separately block the calcium currents, were introduced into the cell (see Chapter 2). Calcium currents could be separately recorded after the replacement of sodium in the extracellular solution by impermeable Tris ions; glutamate anions, which do not affect calcium currents when applied intracellularly, were used as the intracellular ions. The changes in surface potential induced by binding of corresponding ions to the charged groups on the membrane surface have been evaluated from the shifts of the descending branch of the current–voltage curves (I–V curves) for the corresponding ionic currents appearing after replacement of Tris-chloride in the external solution by different concentrations of divalent cations. In this region of membrane potential the amplitude of ionic currents is determined predominantly by the process of potential-dependent activation (opening) of ion channels; for quantitative determination the potential inducing a half-maximal activation can be used. As in all investigated excitable membranes, an increase in the concentration of divalent cations in the external medium induced a shift of the I–V curves for both sodium and calcium currents to the region of more positive membrane potentials. Examples of such shifts are presented in Fig. 1.3. If different divalent cations (Ca^{2+}, Sr^{2+}, Ba^{2+}) were introduced in the same concentrations, the shifts were different, which clearly indicates the different effectiveness of their binding (see the examples given in Fig. 1.4). Together with the shift of the I–V curve, the amplitude of the sodium current increased when the concentration of divalent cations was decreased in the external solution. This effect can also be attributed to the changes in surface potential leading to the accumulation of ions near the membrane surface; during neutralization of anionic charges the accumulation should diminish and the near-membrane concentration should decrease (see below). For quantitative evaluation of the results using the equation (1.5) we are dealing with three variables: surface potential φ, dissociation constant of ion complexes K_d, and charge density σ. The surface potential can be taken as a parameter because in the experiment we measure not its absolute value but only the relative change during alterations in the ionic composition of the external solution inducing the shifts in I–V curves. The dissociation constants and charge densities have been evaluated by successive computer approximation from the experimentally observed shifts at different concentrations of divalent cations. Table 1.1 presents the corresponding values of K_d (mM) and σ_0 (e nm^{-2}) for charged groups that influence the sodium and calcium currents in the membrane of snail neurons.

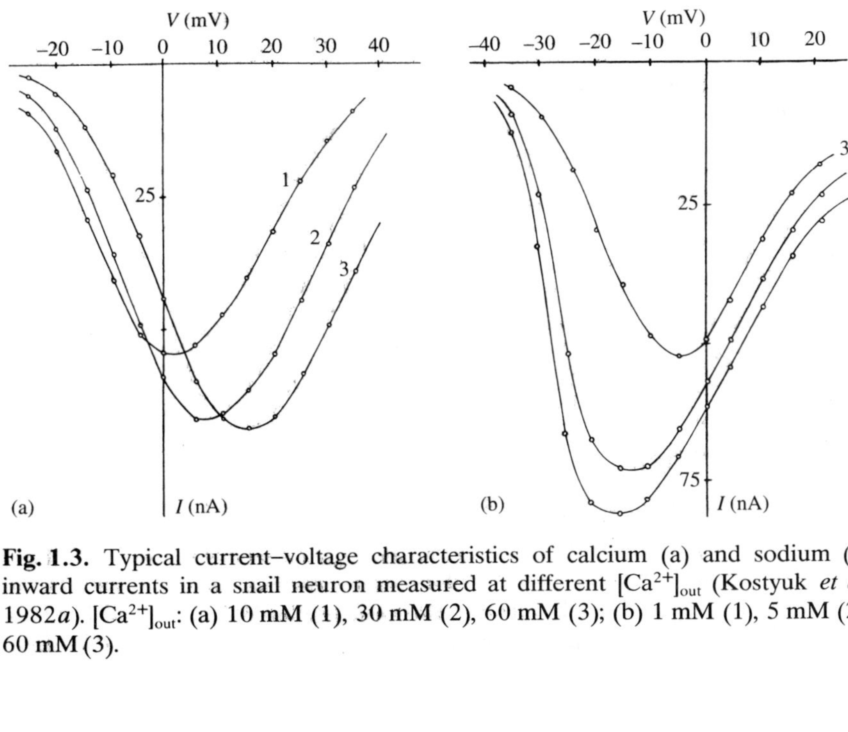

Fig. 1.3. Typical current–voltage characteristics of calcium (a) and sodium (b) inward currents in a snail neuron measured at different $[Ca^{2+}]_{out}$ (Kostyuk *et al.* 1982*a*). $[Ca^{2+}]_{out}$: (a) 10 mM (1), 30 mM (2), 60 mM (3); (b) 1 mM (1), 5 mM (2), 60 mM (3).

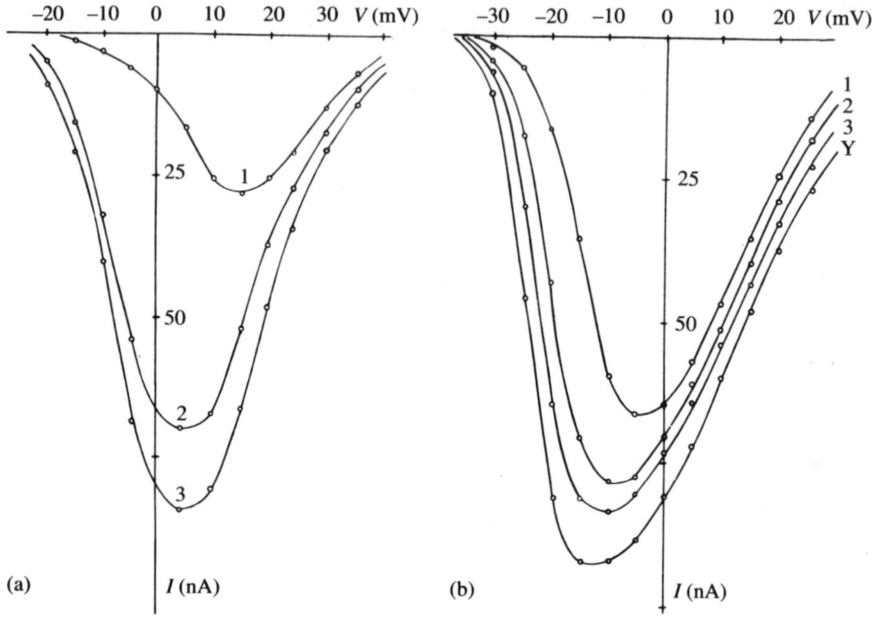

Fig. 1.4. Typical current–voltage characteristics of calcium (a) and sodium (b) currents in a snail neuron measured in the presence of different divalent cations in the external solution (Kostyuk *et al.* 1982*a*). The cations were Ca^{2+} (1), Sr^{2+} (2), Ba^{2+} (3), Mg^{2+} (4). The concentrations used were 10 mM (a) and 5 mM (b).

Table 1.1. Value of K_d and σ_0 for charged groups that influence sodium and calcium currents in the snail neuron membrane

	K_d^{Ca}	K_d^{Sr}	K_d^{Ba}	σ_0
Calcium currents	15 ± 2	50 ± 1	90 ± 2	0.23 ± 0.005
Sodium currents	10 ± 1	17 ± 2	40 ± 1	0.17 ± 0.05

The pH of the solution was 7.7 for sodium currents and 7.0 for calcium currents.

For the membrane of mammalian (rat) neurons the following values have been obtained: $\sigma_0 = 0.15$ e nm^{-2}, $K_d^{Ca} = 100$ mM (Kostyuk and Mironov 1982). As can be seen from the data, the density of charged groups measured from the *I–V* curve shifts for different ionic currents is somewhat different. According to our data, the mean distance between such groups is 20–24 Å. This value is close to that obtained by Mozhayeva and Naumov (1972*a*, *b*) for the frog Ranvier node membrane. However, other authors have described a denser location of the charged groups (11 Å for the axonal membrane of *Myxicola*, Begenisich 1975; 10 Å for the frog Ranvier node, Hille *et al.* 1975; 9 Å for the oocyte membrane, Ohmori and Yoshii 1977).

Our data about the effectiveness of divalent cation binding to nerve cell membranes are somewhat different from those obtained on other tissues. Begenisich (1975) found a K_d value for Ca^{2+} above 10 M, the value given by Ohmori and Yoshii (1977) was between 2 and 5 M, and Mozhayeva and Naumov (1972*a*, *b*) found a value of 200 mM. The reason for such discrepancies seems to be due to the different methods used for the calculation of charge density and dissociation constants. In most cases the density of charges was first calculated from the $I-V$ shifts using only the Gouy–Chapman theory (without taking into account ion binding), and only when there was a deviation from theoretical expectations was a correction for possible binding made. In our approach both values were calculated simultaneously. A comparison of the results indicates that the observed shifts of the $I-V$ curves can be approximated in the framework of Stern's theory by assuming either a high density of surface charges and low binding or a low charge density and high binding (like our results). As it is impossible in such experiments to measure the value of the surface potential directly, an independent criterion could be to use the data about divalent cation binding to similar structural groups in artificial membranes. Hammoudan *et al.* (1981), working on phosphatidyl serine vesicles, obtained values of $K_d^{Ca} = 13$ mM, and $K_d^{Mg} = 50$ mM; Ohki and Kurland (1981) on phosphatidyl serine monolayers found $K_d^{Ca} = 33$ mM and $K_d^{Mg} = 100$ mM. These data confirm our conclusion about a lower charge density and more effective binding of divalent cations to the surface of the neuronal membrane. Similar conclusions can be made from the measurement of surface charges of the neuronal membrane by another technique—electrophoresis of isolated cells.

Experimental determination of calcium ion binding to the neuronal membrane from changes in electrophoretic mobility of the cells

Measurement of changes in the electrophoretic mobility of isolated cells can be used as another indirect method for the evaluation of ion binding to the charged groups on the cell surfaces. Such measurements have been made on sensory neurons from new-born rats, isolated without enzymatic treatment (Dolgaya and Mironov 1984). Glass tubing, with an internal diameter of 3.5 mm and length of 12 cm, having a right-angle opening (2 × 20 mm) in its upper part, has been used. Through the opening the tubing was filled with a suspension of isolated cells in the corresponding solution, and the opening was sealed with a cover glass. The side openings were connected through agar bridges to non-polarizable electrodes passing the electrophoretic current (300 V, 0.5–3 A). The movement of a single cell was observed through a microscope focused on the so-called lower stationary level. The cell's velocity was calculated from the time necessary for it to pass a certain distance (100–200 μm). Data from many cells (50–100) were averaged and statistically treated. To test the experimental procedure, the electrophoretic mobility of erythrocytes had been tested previously; the values obtained corresponded well to those known from the literature.

As predicted, an increase in Ca^{2+} concentration in the solution decreased the electrophoretic mobility of nerve cells. The effect could be explained both by screening of surface charges and by binding of ions to the charged groups. If calcium ions were replaced by an equivalent amount of organic divalent cations of dimethonium (ethan-1,2-*bis*-ethyl-dimethylammonium) or hexamethonium (hexan-1,6-*bis*-trimethylammonium), and EDTA was added for complete removal of calcium ions from the solution, the electrophoretic mobility became less reduced, indicating that these ions exert only a screening effect on their surface charges and do not bind to the charged groups (Alvarez *et al.* 1983). An approximate determination of the dissociation constant could be made from data about the concentration of Ca^{2+} necessary for a twofold decrease in electrophoretic mobility compared with that of an equivalent amount of dimethonium or hexamethonium. The value obtained was 20 mM, which is close to that found from the shifts of I–V curves.

The electrophoretic mobility of nerve cells could also be altered by changing the pH of the solution. Mobility was practically constant during pH changes between 5 and 9, but decreased drastically at pH values below 5. This effect obviously reflects the titration of some anionic groups with pK = 4.2 (the value determined from a twofold decrease in electrophoretic mobility).

The data presented can be compared with those about Ca^{2+} binding on the surface of erythrocytes and lymphocytes, also obtained by electrophoretic measurements (Seaman *et al.* 1969). Two types of acidic groups have been found on the surface of these cells; they react with Ca^{2+} and represent the residues of neuraminic (sialic) acid and carboxylic groups of some membrane proteins. The calculated apparent dissociation constants for Ca^{2+} were within 25–65 mM, similar to those for the neuronal membrane.

The microelectrophoretic technique can also be used for determination of the type of groups responsible for ion binding on the surface of the neuronal membrane if some specific reagents are used. Pre-treatment with trypsin (0.1 per cent) at 22 °C drastically decreased electrophoretic mobility. Pre-treatment with neuraminidase and hyaluronidase (0.3 per cent) did not exert any effect, contrary to the findings on erythrocytes and neuroblastoma cells. *N*-Bromosuccinamide, which affects the carboxylic groups, decreased the mobility, whereas 2,4,6-trinitrosulphonic acid, which affects the amino groups, did not change it substantially. Thus we suggest that the charged groups on the surface of the neuronal membrane are formed mainly by the carboxylic residues of some protein-like molecules and not of the sialic acids. It should be mentioned that other investigators also deny the participation of sialic acids in the construction of the surface membrane of snail neurons (Bolognani *et al.* 1981).

Microelectrophoretic measurements can also answer the question of how divalent cations bind to the charged groups on the membrane surface. At all concentrations of such ions only a decrease in their mobility has been observed; the reversal of direction of movement never occurred. The same has been reported for the electrophoretic mobility of erythrocytes and lymphocytes (Seaman *et al.* 1969). These findings support the earlier suggestion that divalent cations bind to the fixed groups on the cell surface according to the mode MX_z, only decreasing but not reversing the surface potential.

It should be kept in mind that microelectrophoretic measurements reveal not the surface potential itself but the potential in the gliding plane of the particle (ζ potential). If the cellular membrane has an additional envelope in the form of glycoprotein chains ('glycocalyx'), which can also contain fixed charges distributed throughout its volume, the value of ζ potential, according to theoretical considerations, may be about half the value of the true surface potential (Dolgaya and Mironov 1984). Just such a difference between the electrokinetic and surface potential has been observed in the microelectrophoretic experiments.

The thickness of the suggested glycoprotein envelope can be estimated experimentally. The technique suggested by Mironov is based on resonance transfer of excitation energy from the tryptophan residues in the membrane surface to molecules of a fluorescent probe, 2-*n*-toluidine-naphthaline-6-sulphonic acid (TNS), adsorbed in the lipid matrix of the membrane. This transfer is possible only at distances which are less than the so-called Forster critical radius (for the pair tryptophan–TNS this radius is 2.5 nm). From the limiting size of such transfer, obtained after saturation of the lipid bilayer with TNS, the proportion of the tryptophan residues that do not participate in energy transfer can be estimated. It turned out that about 40 per cent of the tryptophan residues are not involved in the process—they are located at a distance from the probe molecules of more than 2.5 nm. If we assume that the tryptophan residues are equally distributed in the glycoprotein envelope the thickness of the latter can be estimated as $d = 2.5/0.6 = 4$ nm. This estimation agrees with data obtained for neuroblastoma cells ($d = 5$ nm).

These findings necessitate a certain reconsideration of the traditional view of the molecular basis of the surface potential of the neuronal membrane. In all previous investigations the surface potential has been analysed on the basis of the Gouy–Chapman theory in its original form, or in modifications considering ion binding at the membrane surface. It has been assumed implicitly that surface charges are distributed in one plane corresponding to the boundary between the lipid matrix and the aqueous phase, and that membrane phospholipids play an important role in the formation of surface charges. This view has to be changed in favour of an important role of charges distributed throughout the volume of the glycoprotein envelope.

Functional meaning of ion binding at the surface of the neuronal membrane

Calcium ions bound at the surface of the cell membrane are obviously indispensable for the maintenance of the structural integrity of the latter; it has been known for a long time that in calcium-free solutions the stability of the membrane decreases. This is manifested, for example, in a progressive decrease in its resistance (increase in leakage). One of the stabilizing factors may be the formation of complexes MX_z, in which Ca^{2+} ions act as tangential crosslinks between individual membrane groups.

Functionally, both the screening effect of calcium and its binding to negatively charged groups may form an effective mechanism for the modulation of membrane processes governed by the transmembrane potential gradient. When the concentration of Ca^{2+} in the extracellular medium increases and the surface charge correspondingly decreases, the intramembrane electric field will increase (see Fig. 1.5). Therefore, a stronger external electric field will be necessary to activate the potential-dependent membrane structures like ion channels and so membrane excitability will be decreased. When the concentration of Ca^{2+} decreases, the functional effects will be opposite. Both effects could certainly take place under physiological conditions in the organism when the level of ionized calcium is shifted in one or other direction.

Because the binding of divalent cations, as has been shown above, is an important component of their interaction with the membrane, one has to keep in mind one more significant functional consequence. The presence of negative fixed charges results in the accumulation of cations near the membrane surface because of their electrostatic attraction. For this reason the effective concentration of such ions important for different membrane functions (for instance, for the generation of a transmembrane current) will be different from that in the bulk solution. After binding of divalent cations (as well as protons) the value of the surface charge will decrease, together with a decrease in the near-membrane concentration of attracted ions. The presence of this effect can be clearly seen in examples already presented in Figs. 1.3 and 1.4. The maximal amplitude of the inward sodium current induced by membrane depolarization depended both on the concentration and type of divalent cations present in the extracellular medium: it decreased with an increase in Ca^{2+} concentration and increased with the replacement of Ca^{2+} by less effectively bound Sr^{2+}, Ba^{2+}, or Mg^{2+} ions. If we assume that the maximal amplitude of the inward sodium current depends linearly on the

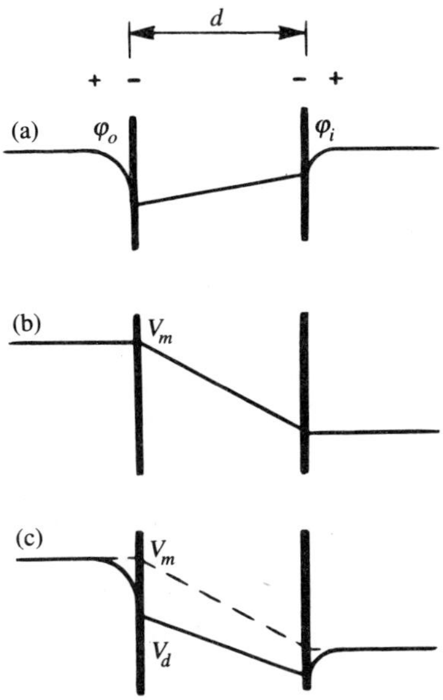

Fig. 1.5. A scheme of potential distribution in the system: external medium–membrane–internal medium (from left to right). Situations are shown with (a) no concentration gradient of the penetrating ion and fixed charges present; (b) fixed charges absent and concentration gradient present; and (c) both fixed charges and concentration gradient present. V_m, potential difference between the media; ϕ_o, external surface potential; ϕ_i, internal surface potential; V_d, total membrane potential difference; d, thickness of the membrane.

concentration of these ions near the membrane surface, then the results obtained can be considered as a good confirmation of this suggestion. The plot in Fig. 1.6 presents these data in the form of a dependence of the logarithm of the sodium current on the surface potential at different concentrations of Ca^{2+} ions in the medium and after their replacement by other divalent cations (the surface potential has been estimated according to eqn 1.4). A similar dependence has been found by Ohki (1978) for sodium currents in the axonal membrane and Ohmori and Yoshii (1977) for sodium and calcium currents in tunicate oocytes. This mechanism has to be taken into account during the analysis of the effects of divalent cations on the electrophysiological characteristics of excitable membranes and of the action on them of different charged pharmacological substances. We shall return to this problem later.

Using the measurements of the electrophoretic mobility of cells, we obtain averaged data about their surface potential. When comparing them to the

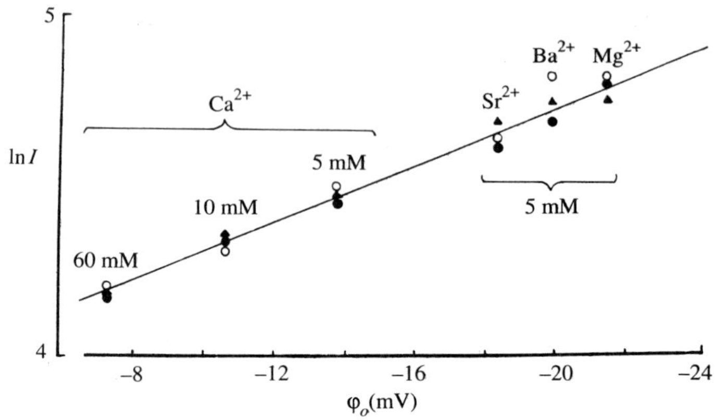

Fig. 1.6. Dependence of the logarithm of sodium current in a snail neuron on the surface potential (Kostyuk *et al.* 1982*b*). Points represent mean values for three cells. The concentrations and types of divalent cation are indicated.

data from $I-V$ curve shifts for different currents, considerable differences often appear. They may indicate an uneven distribution of charged fixed groups near different types of ion channels. Other assumptions can also be of importance. So, for simplicity, it has been assumed that monovalent cations do not bind to the membrane, and exert their effects only by charge screening. But some binding of these ions still takes place (the K_d is about 1.7 M; Eisenberg *et al.* 1979), inducing an error of about 20 per cent. For this reason the suggestion that there is an unequal distribution of charged groups around different types of channels should be considered with caution.

As will be shown later, charged groups capable of binding ions from the external medium may be present directly in the ion channels that cross the membrane; such binding exerts a powerful influence on the function of the channels. However, the presence of such charges can hardly exert any substantial influence on the surface potential of the membrane because of the scarcity of channels in the membrane.

Conclusions

The calcium ions present in the extracellular medium interact with the external surface of the cell membrane. This interaction is due both to electrostatic attraction of ions by fixed anionic groups and to the binding of ions to such groups. Both attraction and binding favour calcium ions rather than sodium ions; for this reason calcium ions, despite their low concentration, can exert a substantial influence on many membrane functions. The effects of electrostatic attraction and binding of calcium and other poly-charged ions overlap in a complicated way; therefore their quantitative description cannot always be precise. However, there are reasons to assume that the effectiveness of ion binding to the membrane anionic groups is higher than usually accepted, and that the corresponding dissociation constants are in the region of tens of millimoles. The use of chemical probes indicates that the nature of charged groups in the membranes of different cells may be quite different; in the neuronal membrane, calcium ions bind mainly to the carboxylic residues of peripheral protein-like molecules.

2. Calcium conductance of the neuronal membrane

History

All living cells maintain, as well as sodium and potassium ion gradients, a steep concentration gradient of calcium ions between the extracellular fluid and the cytoplasm. This gradient is much steeper than those for the two other ions, despite the low concentration of extracellular calcium; this phenomenon is due to the ability of living cells to immobilize calcium ions inside or to expel them, and thus to keep their intracellular activity at an extremely low level (see Chapter 8). For this reason, calcium ions should enter the cell very effectively along this gradient as soon as the cellular membrane becomes permeable to them. However, in the resting state calcium ions obviously do not penetrate through the membrane, and the main pathway for their permeation is formed by specialized macromolecular structures, ion channels, which can be opened for diffusional movement of ions. In 1953 Fatt and Katz showed for the first time that the action potential in crayfish muscle fibres can be evoked in sodium-free extracellular solutions and consequently in the absence of inward sodium currents. Later it was demonstrated that the inward current necessary for membrane depolarization is carried in this case by Ca^{2+}, or even more effectively by Ba^{2+} or Sr^{2+} ions (Fatt and Ginsborg 1958). More extensive data about Ca^{2+} fluxes through the muscle fibre membrane during excitation has been obtained by Hagiwara's group, working on *Balanus* muscle fibres. In this case it was always possible to elicit full-sized calcium-dependent action potentials after artificial decrease of the intracellular calcium level by injection of calcium-chelating substances (Hagiwara and Naka 1964; Hagiwara *et al.* 1964; Hagiwara and Nakajima 1966). The changes in 'calcium' spike amplitude during variations in extracellular calcium concentration were found to be close to those for a theoretical calcium electrode, indicating a high selectivity of the ion-conducting pathways for calcium ions. At the same time, certain difficulties arose when interpreting the data. As 'calcium' spikes appeared when the intracellular ionized calcium was reduced to a very low level (below 8×10^{-8} M), the corresponding electromotive force (EMF) should be very large (about +150 mV), greatly exceeding the observed spike amplitude. A possible explanation could be that, in parallel, the potassium conductance of the membrane was also activated. In fact, if transmembrane currents were directly measured during generation of 'calcium' spikes by the voltage-clamp technique, their equilibrium potentials were found to be much lower than those expected theoretically for a calcium electrode (Hagiwara *et al.* 1974). When special measures were taken to depress potassium conductance, the equilibrium potential shifted to more positive values (Keynes *et al.* 1973).

In 1956 it was detected that the action-potential plateau in mammalian cardiac muscle is much more sensitive to the removal of calcium ions from the extracellular solution than the initial fast component of the spike (Coraboeuf and Otsuka 1956). The same is also true for the frog myocardium (Niedergerke and Orkand 1966). It became clear that in cardiac muscle fibres a considerable influx of Ca^{2+} also occurs during excitation. As is well known, adrenalin considerably prolongs the cardiac action potential; in parallel an increase in calcium influx also occurs (Reuter 1965). The use of the voltage-clamp technique has established the existence of separate systems of sodium and calcium conductances: in sodium-free solution the fast component of the inward current disappeared, but a slow component, dependent on the external calcium concentration, remained (Reuter 1967). Both components could also be separated if Co^{2+}, Ni^{2+}, or Mn^{2+} were introduced in the external solution, as well as the organic substances verapamil or D-600 which specifically depressed the 'calcium' component of the current (Kohlhardt *et al.* 1972*a*, *b*, 1973). The equilibrium potential of the slow current was quite low, about +150 mV. This could be explained by incomplete selectivity of the corresponding ionic pathways in the membrane in passing Na^+ and K^+ ions as well as Ca^{2+}. The relation P_{Ca}/P_{Na} or P_{Ca}/P_K was assumed to be about 1 : 0.01 (Reuter and Scholz 1977*a*).

In smooth muscle fibres the ability to retain excitability in sodium-free solutions for a long time has been shown by Holman (1958), Bülbring and Tomita (1969), and many others. The amplitude of the action potentials in this case depended on the concentration of Ca^{2+} in the external medium; they were insensitive to the sodium-current blocker tetrodotoxin (TTX) but could be inhibited by D-600, Co^{2+}, Ni^{2+}, and Mn^{2+}. These data justified the conclusion that in smooth muscle fibres the calcium conductance is not only an important but probably the main mechanism in the generation of propagated impulses. The presence of calcium inward currents in smooth muscle fibres has been shown directly using voltage-clamp and sucrose-gap techniques (Kumamoto and Horn 1970; Anderson *et al.* 1971; Kao and McCullough 1975; Bury and Shuba 1976); however, a detailed analysis of the calcium currents in such fibres was technically more difficult than in cardiac fibres. Calcium action potentials have also been detected in unicellular organisms (*Paramecium*) in the parts of the membrane adjoining the cilia (Dunlap 1977).

The ability of the neuronal membrane to generate full-sized calcium-dependent action potentials in sodium-free solutions was first shown by Oomura *et al.* (1961) on neurons of the mollusc *Onchidium*. Later on it was shown that the same is true for the neurons of the snail *Helix pomatia* (Gerasimov *et al.* 1964). These findings have been confirmed by several investigators on the neurons from a variety of molluscs (Junge 1967; Geduldig and Junge 1968; Meves 1968). Soon afterwards the presence of 'calcium' action potentials was shown in most neurons investigated: sympathetic neurons of the frog (Koketsu and Nishi 1969), secretory neurons of the crayfish (Iwasaki and Satow 1971), neurons of the guinea-pig Auerbach's plexus (Hirst and Spence 1973), leech Retzius cells (Kleinhaus and Prichard 1975), insect motoneurons (Pitman 1975), motoneurons from the spinal cord of frog (Barret and Barret 1976; Alvarez-Leefmans and Miledi 1980) and cat (Schwindt and Crill 1980), sensory neurons of mice (Ransom and Holz 1977; Matsuda *et al.* 1978) and pigeon (Goerke and Pierau 1980), mammalian and avian cerebellar Purkinje neurons (Llinas and Hess 1976; Llinas and Sugimori 1980*a*, *b*), rabbit hippocampal neurons (Schwartzkroin and Slawsky 1977), cells from the neural part of the pituitary gland (Douglas and Taraskevich 1980), etc. The presence of an inward calcium current in nerve cells has also been detected in an indirect way, by comparing the amplitude of the outward current and the amount of K^+ leaving the cell; the described 'deficit' has been attributed to the presence of a masked inward current (Lux and Eckert 1974).

Thus, the somatic (and obviously dendritic) membrane of a nerve cell seems to be different in principle from the axonal membrane, where the generation of an action potential is always completely stopped after the removal of sodium ions from the external solution. Obviously, even if the axonal membrane includes a calcium conductance, it is so negligible that the calcium transmembrane current cannot produce a detectable charge of the membrane capacitance.

Besides the calcium conductance activated by membrane depolarization, the presence of a similar conductance regulated by the action of some chemical substances on membrane receptors has now been established in many neurons (MacDermott *et al.* 1986; Asher and Nowak 1987). The presence of a specific calcium conductance has also been found in the membrane of certain cytoplasmic organelles. Among the best known is the sarcoplasmic reticulum of striated muscle fibres, which is the source of calcium ions necessary for the triggering of muscle contraction (Imagava *et al.* 1987*a*, *b*; Fill and Coronado 1988); recently the same property has been detected in the endoplasmic reticulum of nerve cells.

Characteristics of transmembrane calcium currents

Direct data about the movement of calcium ions through the neuronal membrane could be obtained by the measurements of transmembrane electric currents using the voltage-clamp technique or by direct determinations of the changes in the intracellular calcium level using special reagents. During electrophysiological determination of calcium currents, the main problems are to obtain a reliable fixation of the membrane potential on the whole surface of the cell and a complete separation of this current from the currents carried by other ions. The first problem can be solved most successfully by using cells having a simple spherical (unipolar) form. Isolation of such cells is valuable because it removes the single existing neurite; the hole in the membrane seals itself quickly and the whole cellular surface becomes completely controllable. The isolation also removes the diffusional barriers which may hinder access to the external surface of the membrane. Such isolation has been initially for snail neurons (Kostyuk *et al.* 1974*a*). Other authors also used ligation of the axon near the soma (Standen 1975*a*; Connor 1977, 1979; Tillotson and Horn 1978; Adams and Gage 1979*a*). For elongated cells (muscle fibres) the sucrose-gap technique, or sometimes the introduction of several intracellular microelectrodes, has been used.

The separation of calcium currents from sodium currents can be achieved in most cases by the use of specific blockers of sodium channels such as tetrodotoxin (TTX). However, in some cases the sodium channels are quite resistant to TTX (mollusc neurons), or special TTX-insensitive sodium channels exist besides the usual TTX-sensitive ones (mammalian dorsal root ganglion neurons). However, in these cases the calcium current can also be subtracted from the total inward current by replacing the sodium ions in the extracellular medium by some impermeable ions; its current–voltage characteristic is always quite different from that of sodium currents (Geduldig and Gruener 1970; Krishtal and Magura 1970; Kostyuk *et al.* 1974*b*; Standen 1975*a*, *b*; Eckert and Lux 1976). An example of separation of the calcium and sodium inward currents in a snail neuron has already been shown in Fig. 1.3.

However, the oppositely directed potassium currents, which due to their delayed activation strongly mask the calcium currents, cannot be completely suppressed by pharmacological tools (tetraethylammonium, 4-aminopyridine). Obviously, the most effective way for their elimination would be the removal of charge carriers from inside the cells. This has been done successfully on dialysed neurons with the use of the techniques described above, first on snail neurons (Kostyuk *et al.* 1975; 1977; Kostyuk and Krishtal 1977*a*,*b*) and later on isolated amphibian (Veselovsky *et al.* 1977) and mammalian (Kostyuk *et al.* 1981*c*, *e*, *f*) sensory neurons, as well as on clonal neuroblastoma cells (Kostyuk *et al.* 1978). Modifications of intracellular perfusion techniques for this purpose have also been used by other groups (Lee *et al.* 1978; Takahashi and Yoshii 1978). The modification proposed by the latter authors has been used successfully for the separation of calcium currents in the membrane of oocytes (see Chapter 7).

Calcium currents can be recorded in snail neurons for several minutes and in mammalian cells for several tens of seconds after the beginning of cell dialysis within a large range of membrane potentials. Nevertheless, with testing potentials above +40 mV an outward current still remains which hinders the investigation of the inward current in this potential region and may even be considered as its reversal. Potassium channel blockers do not affect this outward current. A characteristic property of this 'non-specific' current is its high sensitivity to pH changes in both intra- and extracellular solutions; it can be largely depressed or even abolished by lowering pH_{out} (Doroshenko *et al.* 1978) or elevating pH_{in} (Doroshenko *et al.* 1984*a*). These findings may indicate that the 'non-specific' outward current is produced mainly by the flux of protons through some ion channels which, judging from the kinetics, might be through the channels of the calcium-dependent potassium conductance; however, special proton channels cannot be excluded.

Calcium ions as charge barriers can be replaced only with Sr^{2+} or Ba^{2+}; the corresponding current–voltage characteristics for a snail neuron have already been shown in Fig. 1.4. If they are measured in a really pure form without superposition of outward currents, they always show a high degree of non-linearity: with high positive membrane potential levels the current amplitudes approach zero level approximately exponentially. This type of potential-dependence has to be expected because of a very low intracellular level of charge-carrying ions, and in accordance with the simplest assumption about the constancy of the intramembrane electric field:

$$I_{Ca} = P_{Ca}\left\{\frac{4F^2E}{RT}\right\}\left\{\frac{[Ca]_{out} - [Ca]_{in}\exp(2EF/RT)}{1 - \exp(2EF/RT)}\right\} \tag{2.1}$$

where $[Ca]_{in}$ and $[Ca]_{out}$ are the concentrations (activities) of Ca^{2+} ions inside and outside the cell; I_{Ca} is the calcium current; P_{Ca} is the calcium permeability; and E, F, R, and T are the usual nominations. In natural conditions the level of free calcium ions in the cytoplasm is of the order of 10^{-8} M (see Chapter 8), and so the equilibrium potential for calcium currents should be in the region of +200 mV. However, in experimental conditions it is practically impossible to shift the membrane potential to such high positive values without killing the cell. One could expect a more linear I–V relation in the case of an artificial increase of the intracellular calcium level, which may be quite simple in the case of intracellular dialysis or perfusion using injections of certain calcium buffers. However, it appears that elevation of this level even to 10^{-7} M results in a reversible depression of calcium currents (Kostyuk and Krishtal 1977*b*). The same effect has been observed by Hagiwara and Naka (1964) after injection of Ca^{2+} ions into *Balanus* muscle fibre. The mechanism of such a blockade will be considered later.

Another way to analyse calcium conductance is based on the measurements of the concentration characteristics of membrane currents using different charge carriers. Such measurements made in several laboratories (Hagiwara 1973; Akaike *et al.* 1978; Valeyev 1979; Ponomarev *et al.* 1980) have demonstrated that these characteristics show definite saturation and can be approximated by Langmuir's isotherm. For this reason they can be considered, according to be Hagiwara's suggestion, as an indication of binding of the penetrating ion to some binding group inside the channel. Some investigators did not observe definite dependence of such binding on the membrane potential (Naruševicius and Rapoport 1982); in our experiments such dependence has always been seen. One should also keep in mind certain difficulties connected with the presence of negative surface charges on the neuronal membrane and the consequential preferable attraction of divalent cations; for this reason their actual near-membrane concentration may differ substantially from that in the bulk solution. All these factors have been taken into account in our calculations, as already described in Chapter 1, and the pK values ($-\log K_d$) so obtained for the suggested complexes of penetrating ions with the channel binding site are presented in Table 2.1. This table also includes data about the effects on calcium currents of other divalent cations which do not penetrate through the channels but, on the contrary, block their conductance. This blocking effect can also be described by Langmuir's isotherm and is obviously due to competitive interaction at the same binding site. In addition, the table presents pK values for binding of divalent cations with a carboxylic group (glycine) in aqueous solution.

From these data a conclusion can be made that divalent cations form a continuous sequence according to their influence on membrane conductance, which coincides with the sequence of their binding to a carboxylic group. All ions that bind weaker than Ca^2 pass through the membrane; the easier, the weaker they bind. All ions with stronger binding block the conductance, again in relation to the effectiveness of their binding. Some of the blocking ions that completely abolish the calcium currents, can induce small transmembrane currents *per se*; for example, Akaike *et al.* (1983) observed measurable Mn^{2+} and Zn^{2+} currents in snail neurons.

A similar model has been proposed for the calcium conductance of brain synaptosomes by Nachshen (1984). According to this model, the binding centre in the calcium channel consists of two negative charges separated by a 2 Å distance. The distance between oxygen atoms in a carboxylic group is about the same (2.1 Å). Because of the effect of delocalization of electron densities, each of the oxygen atoms here carries a charge equal to 0.5 e, which is half as much as in Nachshen's model. Thus in the two models the interaction energy of the penetrating cation with the channel must differ by about a factor of two, but their relative affinity would be identical. From a physical point of view both models are identical; from a chemical one ours seems preferable because it offers for the binding centre a real molecular structure.

The results presented indicate that the ability of the neuronal membrane to select and pass divalent cations is based on mechanisms which are in principle different from those for the sodium conductance. The sodium channels select penetrating ions mainly by their steric features (see Hille

Table 2.1. Divalent ions binding in calcium channel: pK values

Object	Ion					Authors
	Ba^{2+}	Ca^{2+}	Co^{2+}	Ni^{2+}	Cd^{2+}	
Snail neuron	1.0	2.0	3.2	3.5	4.3	Kostyuk *et al* (1982*a*)
Rat sensory neuron	0.6	1.3	2.5	2.6	4.3	Kostyuk and Mironov (1982)
Carboxylic group	0.8	1.4	3.2	2.2	4.8	Martell and Smith (1977)

1975); on the other hand, calcium conductance is based mainly on ion binding which forms a necessary step in the penetration of ions through the membrane. However, this binding should not be too strong otherwise it would block the channels.

It should be mentioned that attempts have been made to reject the idea about ion binding in calcium channels and to relate all changes appearing in membrane currents during variations of the divalent ion concentration in the external solution, or their replacements, only to the changes in the near-membrane concentration of these ions (Wilson *et al.* 1983). This approach does not agree with our results and those of many other investigators. For instance, Cota and Stefani (1984) performed careful calculations of the changes in the surface charges and near-membrane ion concentrations in striated muscle fibre and concluded that Ca^{2+} ions bind in the channel with $K_d = 0.016$ M and Ba^{2+} with $K_d = 0.095$ M, values that are very close to our data about calcium channels in the neuronal membrane. They also consider a carboxylic group as the most probable site for such binding.

Calcium currents in the somatic membrane are so far the best analysed, and we also have detailed information about the function of the corresponding individual ion channels (see Chapter 3). The somatic membrane differs markedly in this respect from the axonal membrane, in which the penetration of Ca^{2+} ions can hardly be demonstrated. Nevertheless, using very sensitive techniques, the influx of Ca^{2+} in the axon during generation of an action potential can just be detected. Hodgkin and Keynes (1957), working with ^{45}Ca, have shown that about 0.006 pM cm^{-2} of Ca^{2+} enters the axon during the spike. This amount is negligible compared with the amount of Na^+ entering through the same membrane surface (4 pM cm^{-2}). With the help of the calcium-sensitive protein, aequorin, the sensitivity of measurements has been substantially increased, allowing separation of the calcium influx into two phases. The initial phase coincided with the influx of Na^+ and could be blocked by TTX; it represents the transfer of Ca^{2+} through the 'sodium' pathways due to their imperfect selectivity (Baker *et al.* 1971). The relation, P_{Ca}/P_{Na}, for this pathway was about 1:100 (Meves 1975). The second slow phase of the Ca^{2+}entry was insensitive to TTX but could be blocked by Mg^{2+}, Mn^{2+}, and Co^{2+}, as well as by the organic calcium-channel blocker D-600. It could be connected with the presence in the axonal membrane of some specific calcium conductance (Baker et al. 1973*a*, *b*). Recently, this phase of calcium entry has been studied on perfused squid axons in conditions of potassium-free extra- and intracellular solutions. The 'late' calcium current could be recorded directly; it could be activated at membrane depolarizations of up to −40 mV and reached a maximum at 0 mV. Barium ions could replace Ca^{2+} as a charge carrier in this current (DiPolo *et al.* 1983).

Thus, although the presence of voltage-operated calcium conductance is a general property of the neuronal membrane, its manifestation is very uneven in different parts of the neuron. Being negligible in the axon, it becomes very prominent in the membrane of synaptic terminals, as was first shown in the experiments of Katz and Miledi (1976*a*, *b*, 1970). Direct recording of calcium currents in axon terminals was difficult because of their small size; but in giant synapses of the squid stellate ganglion, in which two intracellular microelectrodes can be inserted, the presence of a specific calcium current has been proved both electrophysiologically and with the use of aequorin (Llinas *et al.* 1972, 1976).

The characteristics of calcium currents in other excitable tissues are very similar to those already described. This has been shown on cardiomyocytes (Trautwein *et al.* 1975; Reuter and Scholz 1977*a*; Isenberg and Kloeckner 1982), striated muscle fibres of frog (Sanchez and Stefani 1978, 1983) and rat (Donaldson and Beam 1983), oocytes (Okamoto *et al.* 1976*a*, 1977), and sensory hair cells (Ohmori 1984). In cardiomyocytes a specific feature of calcium currents has been observed that is not found in other cells. Working on cells isolated with a glass micropipette, Lee and Tsien (1982) claimed to observe a true reversal of the current through calcium channels due to outward movement of intracellular potassium ions. The reversed current could be blocked by specific calcium-channel blockers (D-600) without changing the reversal potential. Cs^+ ions could also carry the reversed current in case of the absence of Ca^{2+} in the extracellular solution; otherwise the latter were bound and the currents blocked. Perhaps these data indicate some specific features of the 'selectivity filter' of calcium channels in the membrane of cardiac cells. This phenomenon should be studied more carefully in other cells.

Energy profile of the calcium channel

One of the ways to characterize the mechanism of membrane calcium conductance is the construction of energy profile diagrams for the corresponding calcium channels. This approach is based on the theory of absolute reaction rates proposed by Eyring *et al.* (1949); it has been used already by Markin and Chizmadjev (1974), Hille (1975), and others for the description of sodium and potassium channels in cell membranes. The penetration of an ion through the ion-conducting channel is considered as a sequence of binding reactions in energy 'wells' separated by energy barriers which the ion has to overcome. The transition through the *i*-barrier can be characterized by the rate constants k_i and k_{-i}. When constructing the model, it was suggested that there is a certain number of wells and barriers and corresponding forward and backward reactions occur. Using these variables, equations are constructed for the stationary current through the channel. Because of the complexity, these equations are not presented here; they can be found in the papers of Markin and Chizmadjev (1974), Kostyuk *et al.* (1981*d*, 1982*a*). The parameters of the equations are then adjusted by minimizing the least squares deviations of the theoretical $I-V$ curves from the experimentally measured ones for different concentrations of carrier-ions. The parameters so obtained can be used later for the prediction of current characteristics in conditions of as yet untested external solutions or membrane potential shifts.

The use of the above approach for the calcium channels is complicated by the accumulation of divalent cations near the membrane surface already mentioned. Therefore a correction is always necessary. Figure 2.1 shows examples of the $I-V$ curves for calcium and barium currents in snail neurons after such correction, eliminating their shifts due to changes in surface

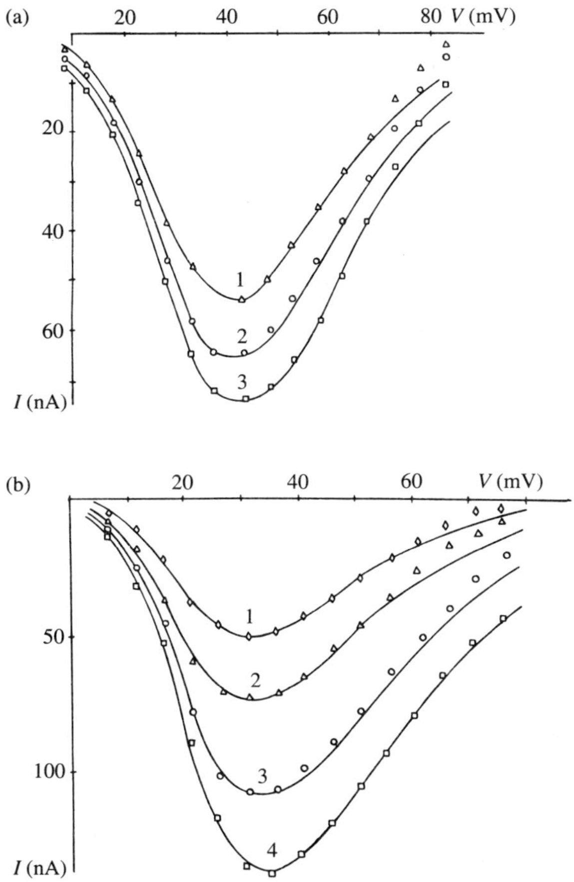

Fig. 2.1. Current–voltage characteristics (experimental points and theoretical curves) of a calcium (a) and barium (b) current in a snail neuron corrected for the surface potential value (Kostyuk *et al.* 1982*a*). The theoretical curve was calculated according to a three-barrier model. $[Ca^{2+}]_{out}$: 10 mM (1), 30 mM (2), 60 mM (3). $[Ba^{2+}]_{out}$: 5 mM (1), 10 mM (2), 30 mM (3), 60 mM (4).

potential. Such curves can be used directly for the determination of model parameters, and our calculations have shown that the ionic current through the calcium channel can be reasonably described in the framework of a three-barrier model presented diagrammatically in Fig. 2.2 (Kostyuk *et al.*, 1981*d*, 1982*a*). The standard deviation of the parameters of such a model from the experimental data obtained on five snail neurons did not exceed 10 per cent. The first potential 'well' obviously corresponds to the binding site for divalent cations, which determines the saturation of currents at high concentrations of the carriers. It should be located at about one-third of the way from the external mouth of the channel, as also follows from the determination of the potential-dependence of binding. The binding characteristics (or the depth of the 'well' in the terms of the model) are the factors that determine the selectivity of the calcium channel. Less effective binding of Ba^{2+} (or a less deep 'well') determines a stronger barium current through the calcium channel.

It has been suggested that ion binding at this site is due to the presence here of a carboxyl group accessible for external ions. Additional characteristics of this process can be obtained from the effects on these currents exerted by lowering the pH of the external solution. Protonation of this group should depress the transition of divalent cations through the channel. Figure 2.3 shows the changes of barium currents in a snail neuron induced by pH decrease; the same figure also presents theoretical curves based on the assumption of pK = 5.8 for this group. Experimental and theoretical curves fit quite well except for small depolarizations; possibly in this case the rate of

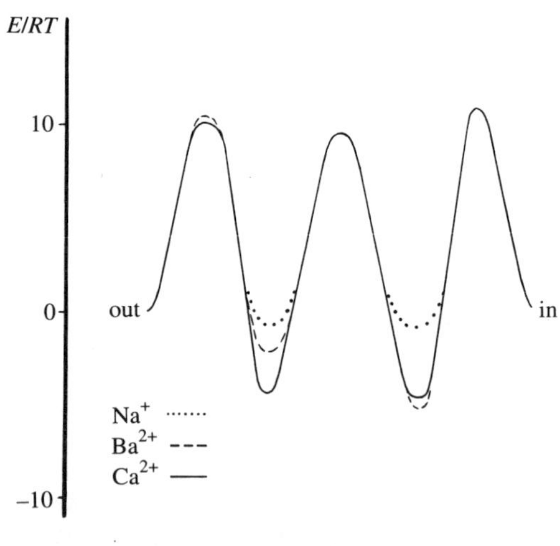

Fig. 2.2 The predicted energy profile of a calcium channel for Ca^{2+}, Ba^{2+}, and Na^{+} ions (Kostyuk *et al.* 1982*a*). The energy of the ion in the channel is expressed in E/RT values.

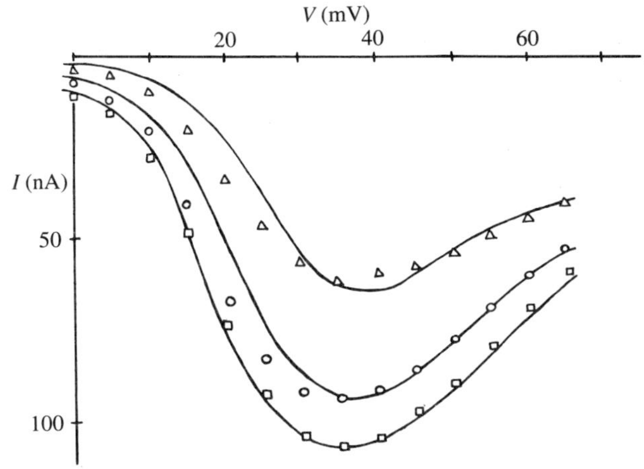

Fig. 2.3. Current–voltage characteristics of a barium current (experimental points and theoretical curves) in a snail neuron at three different pH values of the external solution (Kostyuk *et al.* 1982*a*). pH_{out}: 5.6 (1), 6.2 (2), 7.0 (3). $[Ba^{2+}]_{out}$: 30 mM.

channel opening becomes comparable with the reaction rate between protons and the carboxyl group, and their blocking action cannot fully develop.

The pK value obtained for this binding site is definitely higher than the pK for carboxyl groups in aqueous solution. Probably, this discrepancy should not be considered as an objection against the hypothesis, because the dissociation constants for complexes of carboxyl groups in aqueous solutions and inside the membrane could be different due to the differences in dielectric constants of the medium, influencing mutual attraction of charged elements. The dielectric constant for lipids is about 2 in contrast to 80 for water; for proteins it would have an intermediate value.

Figure 2.4 presents in a graphic form the binding of di- and trivalent cations and protons to a carboxyl group in aqueous solution; it can be seen that the pKs of such binding depend linearly on the inverse sum of the ionic radii of the cation (r_{ion^+}) and the carboxyl group (rP_{COO^-}). In the same figure the second plot presents experimental data concerning the binding pKs for divalent cations and protons in the calcium channel. The relation in this case is also linear but much steeper, corresponding to a dielectric constant of about 20. Such high values of dielectric constant would be expected for protein macromolecules (Mironov 1983*b*).

The proposed model for the energy profile of the calcium channel also predicts another potential well located close to the internal mouth of the channel. One might expect that this well would be more accessible to

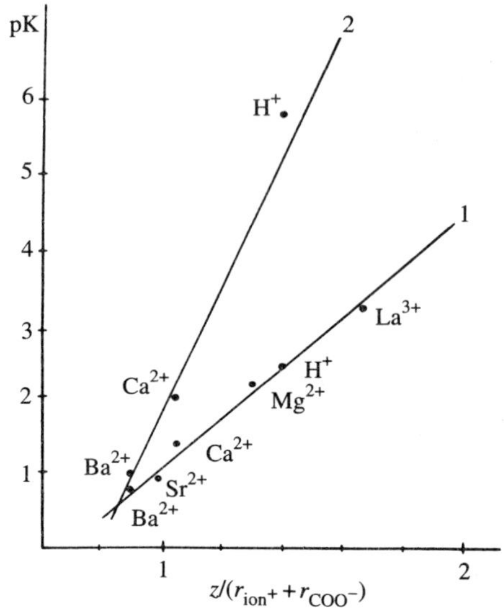

Fig. 2.4. Dependence of the pK values for the binding of different cations with the carboxyl group on the value of ionic radius, calculated (1) for aqueous solution (from Martell and Smith 1977) and (2) for the calcium channel (Kostyuk *et al.* 1982*a*).

influence from the cell interior and so could be defined using the changes in intracellular concentration of divalent cations. As already mentioned, an increase in the intracellular concentration of Ca^{2+} induces substantial changes in calcium currents; they already become depressed by an increase in $[Ca^{2+}]_i$ to a level of 10^{-7} M. If this effect is due to ion binding in the suggested second energy well, this binding would be extremely effective (pK about 8); however, the model predicts pK $\cong$ 2.7. Theoretically the presence of a very effective binding site as the ion moves through the channel would predict a negligible electric conductance of the latter (10^{-15}–10^{-16} A); in fact it is several orders higher (see Chapter 3). For all these reasons it is more probable that the blocking effect of intracellular Ca^{2+} on calcium currents is not due to ion binding in a second energy well in the channel; possibly these ions exert a more complicated effect on the channel-forming macromolecule.

Quite recently a somewhat similar conclusion has been reached from another standpoint. Measurements of barium currents through single calcium channels (see Chapter 3) in guinea-pig cardiomyoctes have shown that their dependence on external Ba^{2+} activity is more complicated than that predicted by simple Michaelis–Menten formalism. This deviation seems to be better explained by the suggestion about the presence of multiple successive divalent ion-binding sites in the channel (Yue and Marban 1990).

Why don't sodium ions pass through the calcium channel?

This question is very important for the understanding of the mechanism of specificity of calcium channels, as Na^+ ions are close in size to permeable Ca^{2+} ions. A clue for solving this question may be found in the discovery of the reversible transition of calcium channels into sodium ones after complete elimination of divalent cations from the extracellular solution by the addition of calcium-chelating substances (Kostyuk and Krishtal 1977*b*). This phenomenon is illustrated in Fig. 2.5. The inward current was recorded from a snail neuron in which its sodium component was practically absent. Then the external normal saline solution was changed for a solution containing the same amount of Na^+ but lacking Ca^{2+} because of the addition of EDTA. Immediately after the change the inward current increased markedly and its I–V curve shifted to high negative potentials (activation threshold about −90 mV). The induced current was definitely carried by Na^+: if these ions were introduced in the cell in the same concentration as outside, the current reversed its direction at membrane potentials close to zero. At the same time, the pharmacological characteristics of the induced sodium current remained similar to those of the initial one: it could be blocked by organic calcium-channel blockers (D-600, nifedipine) even more effectively than the original calcium current, as can be seen from the dose–response curves presented in Fig. 2.6.

The induced current could also be blocked by intracellular perfusion with fluoride solutions, just like the calcium current, and by elevation of the

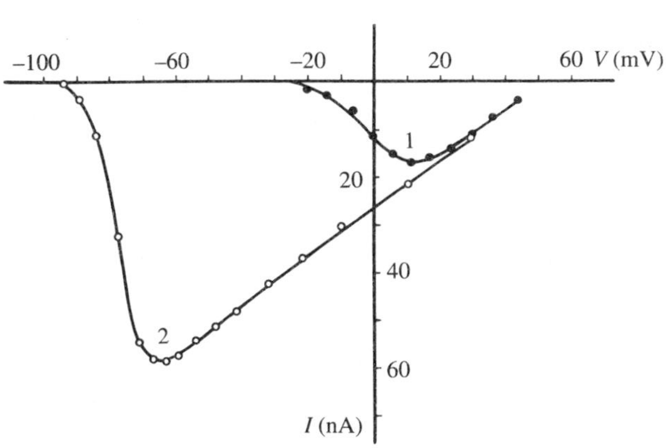

Fig. 2.5. Currents obtained in a snail neuron in Na-free solution with 30 mM $CaCl_2$ (1) and in Ca-free EDTA-containing solution with 30 mM NaCl (2) (Kostyuk *et al.* 1983).

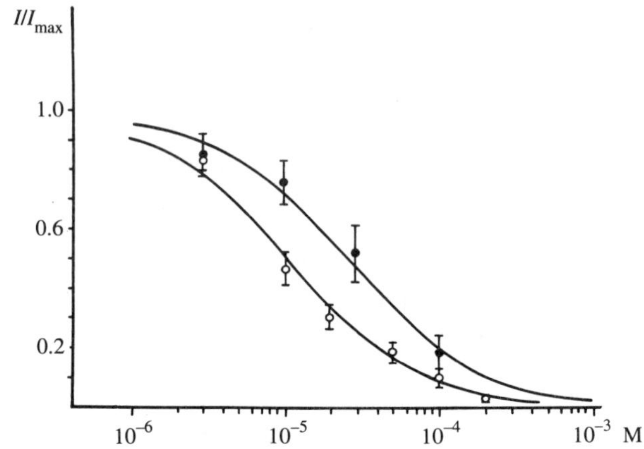

Fig. 2.6. Dose–effect dependence for the blocking action of D-600 on the calcium (●) and induced sodium (○) currents (Kostyuk *et al.* 1983). Measurements on four snail neurons are presented. The solid lines represent Langmuir's isotherms with $K_d = 2.6 \times 10^{-5}$ M and 10^{-5} M, respectively.

intracellular calcium level; such elevation up to 10^{-6} M has no effect on sodium currents flowing through 'natural' sodium channels. Finally, determination of the selectivity sequence for the induced current by measuring the shifts in reversal potential after replacement of different monovalent cations has shown a high selectivity for Na^+, but not one identical to that of 'natural' channels (Kostyuk and Shuba 1982; Table 2.2).

All these findings agree with the conclusion that a profound modification of calcium channels occurs in the absence of divalent cations: they become highly permeable for monovalent cations like 'natural' sodium channels, at the same time retaining some features (not related to selectivity) of calcium channels. A similar modification of calcium channels was later described for

Table 2.2. Monovalent cation selectivity of natural and induced Na^+ currents

Sodium current	Selectivity		
	P_{Li}/P_{Na}	$P_{N_2H_5}/P_{Na}$	P_{NH_3OH}/P_{Na}
Natural	0.80 ± 0.15	0.55 ± 0.07	0.21 ± 0.02
Induced	1.04 ± 0.14	0.44 ± 0.03	0.19 ± 0.02

smooth (Curtis and Prosser 1977), cardiac (Matsuda 1986), and skeletal (Palade and Almers 1978) muscle fibres.

Modification of calcium channels can also be induced by lowering the external pH (Konnerth *et al.* 1987). It is suggested that protons can expel Ca^{2+} from some calcium-binding centres important for the maintenance of normal selectivity of calcium channels (Morad and Lux 1987; Davies *et al.* 1988). This effect may be identical to the earlier described proton-induced sodium conductance in mammalian sensory neurons (Krishtal and Pidoplichko 1981); however, further experiments are necessary to prove it.

For a more detailed analysis of the processes occurring in the calcium channel during transformation, dose–response curves have been obtained for the recovery of channels into calcium-conducting mode after readmission of divalent cations to the extracellular solution, using different buffer systems for varying the level of such ions between 10^{-7} and 10^{-4} M (Kostyuk *et al.* 1983). The concentration dependence of recovery could be approximated by Langmuir's isotherm with distinct dissociation constants for different ions. Examples of such curves are shown in Fig. 2.7. The mean K_d values for Ca^{2+}, Sr^{2+}, Ba^{2+}, and Mg^{2+} were 2×10^{-7}, 3.5×10^{-6}, 1.4×10^{-5}, and 6×10^{-5} M, or, in pK values, $pK_{Ca}:pK_{Sr}:pK_{Ba}:pK_{Mg} = 6.6:5.4:4.8:4.2$. The binding demonstrated only weak potential-dependence.

The values obtained are several orders higher than those already given for binding of penetrating ions in the channel; the ion sequence is also different. They resemble those for specific binding of divalent cations by chelators or calcium-binding proteins (parvalbumin, troponin C, calmodulin). In these cases the high effectiveness of binding is due to the presence of several carboxyl groups in the active centre which form a cage for the bound ion.

We considered these data as an indication of the presence in the calcium channel of another binding site which effectively chelates divalent cations. In normal conditions the Ca^{2+} ion is always present at this site, which keeps the channel in a calcium-selective mode. When this ion is taken away, the channel undergoes conformational changes and starts to pass ions in the same way as the 'natural' sodium channel. We still do not know the intimate nature of such changes. The suggested binding site can be designated as a 'regulatory centre' of the calcium channel. This idea can be illustrated by a diagram suggested by Kostyuk and Mironov (1986) and presented in Fig. 2.8.

It has been already stated that it is very difficult to determine the location of a very effective binding site in the pathway of ions through the channel; the presence of such a deep energy well would result in negligible electric conductance of the channel. The weak potential-dependence of binding also suggests a different location of the corresponding site. The extremely strong shift of the *I–V* curve of the inward current during modification also needs special explanation. As can be seen from Fig. 2.5, it amounts to 60–70 mV; *I–V* curves for sodium currents through 'natural' channels are shifted under similar conditions only by 25–30 mV. Obviously, such a huge difference

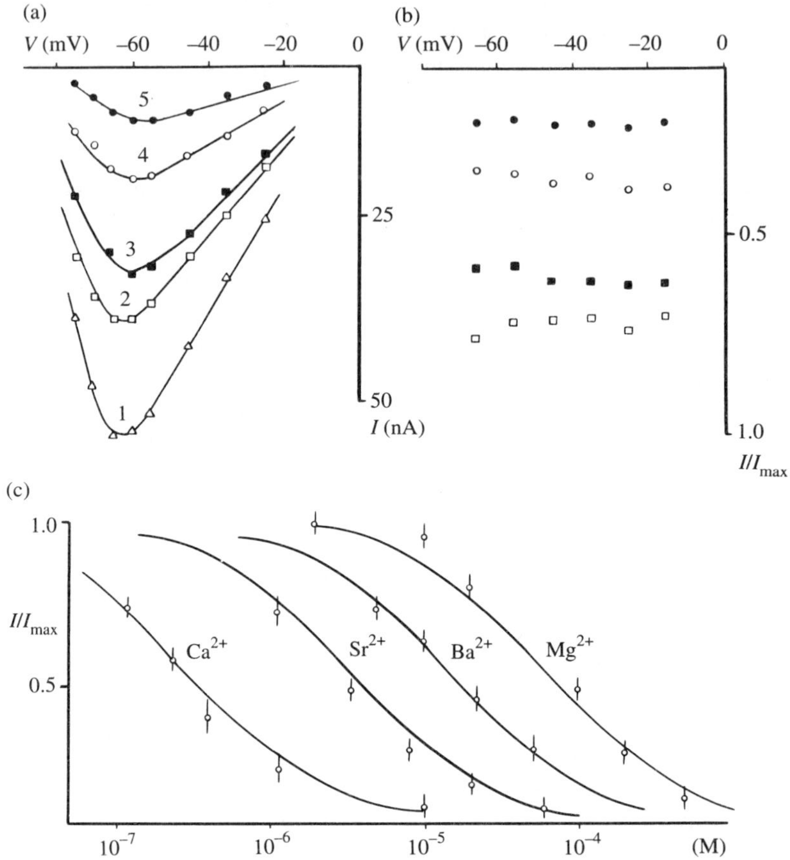

Fig. 2.7. Induced sodium currents in snail neurons at different concentrations of $[Ca^{2+}]_{out}$ (Kostyuk *et al.* 1983). (a) Current–voltage curves obtained at $[Ca^{2+}]_{out}$ 3×10^{-8} M (1), 1.2×10^{-7} M (2), 2.4×10^{-7} M (3), 4×10^{-7} M (4), 1.2×10^{-6} M (5). (b) Relative reduction of the induced sodium current with the increase of external Ca^{2+} concentration depending on membrane potential. The symbols correspond to those used in (a) The current value at $[Ca^{2+}]_{out} = 3.10^{-8}$ M is taken as unity. (c) Dose–effect dependences for the blocking effect of different divalent cations on the induced sodium current. Circles, mean values for five cells; solid lines, Langmuir's isotherms.

cannot be explained only by the changes in surface potential after elimination of divalent cations, as has been done previously. Probably, in the region of each calcium channel an additional strong local electric field appears, which affects the transmembrane potential acting on this channel. Perhaps this local field is generated just by those anionic groups that are unmasked in the 'regulatory centre' after the elimination of the Ca^{2+} ion.

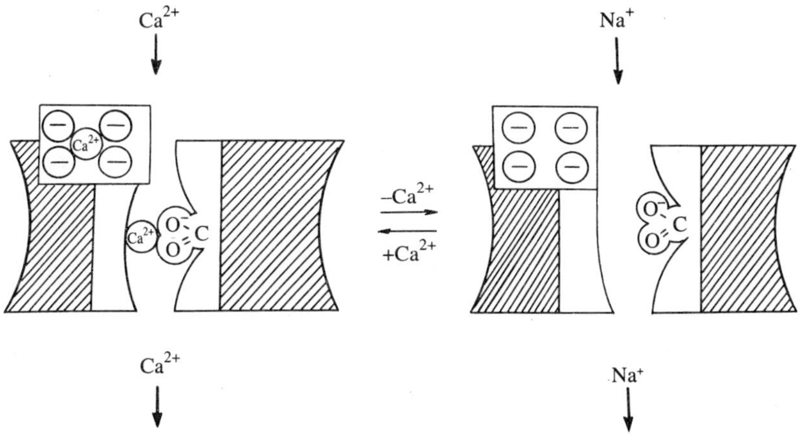

Fig. 2.8. Suggested calcium-binding sites in the calcium channel and their participation in its functioning.

Another approach has been used by Almers *et al.* (1984), Almers and McCleskey (1984), McCleskey and Almers (1985) who tried to reconcile our data about high-affinity binding of divalent cations by calcium channels with the possibility of the location of the binding site directly in the channel. Analysing the concentration dependence of the modifying effect of these ions on the calcium channel in the sarcolemma of frog skeletal muscle fibres, they obtained very similar values ($pK_{Ca} = 6.2$). They suggested that two highly effective binding sites are located *in the channel* close to each other. If one of the sites is occupied, the affinity of the next one decreases because of mutual repulsion of ions. Such repulsion at a distance between the sites of 1.15 nm may decrease the probability of their simultaneous occupation by 12 times (at the dielectric constant of water) or 20 000 times (at the probable dielectric constant in the membrane which may be, as already mentioned, about 20). The repulsion should greatly increase the rate of liberation of divalent cations from the channel compared with that from a channel with a single binding site. At the same time, it should support the selectivity of the channel for divalent cations, and only in case of their complete elimination from the medium (and the channel) would the channel start to pass the weakly binding monovalent cations. The filling of only one site would stop such a non-specific conductance. Current saturation at high extracellular concentrations of carriers ions might be explained by complete filling of both sites. This model also explains a peculiar dependence of the complex calcium + barium current on the concentration of Ca^{2+} in the medium. According to a single-site model, the amplitude of this current should decrease with an increase in Ca^{2+} concentration; in fact the concentration dependence passes through a minimum as predicted by the double-site model. Similar data have also been obtained on cardiac muscle fibres (Hess and Tsien 1984; Hess *et al.* 1986).

The second model is quite attractive because of the simplicity of explanation of several basic features of calcium channels. At the same time it has also weak points. First of all, the calculations of single channel conductance on the basis of this model seem to be quite unrealistic. The accordance with experimentally obtained values after assuming the main role of mutual repulsion of penetrating ions can be obtained only when the height of the last barrier is lowered to a value of about $3RT$. This is lower than the diffusional barrier for Ca^{2+} in the external medium. For all examined channels $E_{\text{barrier}} > E_{\text{diffusion}}$ and it is very difficult to understand the physical meaning of an opposite relation.

The presence of definite conformational changes in the channel-forming macromolecule during its modification has been confirmed by other observations, among them the changes in pharmacological sensitivity already mentioned. Recently, such changes have also been observed in vertebrate neurons for the blocking action of ω-conotoxin (Carbone and Lux 1988). The question of conformational changes of calcium channels during their modification is discussed in more detail by Kostyuk and Mironov (1986) and Morad *et al.* (1988).

Toxic effect of penetrating ions on the calcium channel

Until now we have assumed that divalent cations interact with certain groups in the channel without producing any irreversible changes of the latter. This assumption is made in most investigations of calcium channels, although it is not quite correct. In the early studies of action potentials generated by different divalent cations it had already been observed that 'barium' spikes are quite unstable and diminish rapidly during repetition (Magura 1977). This effect can also be seen during direct recording of barium currents carried through calcium channels in snail neurons (Kostyuk *et al.* 1985*a*). In contrast to calcium currents, barium currents diminished rapidly during repetitive stimulation; the decline slowed down after the decrease of current amplitude to 10–30 per cent of the initial level. The residual current remained quite stable for a long time. Pre-incubation of the cell in barium solution without stimulation did not affect the current. This diminution of barium currents could not be influenced by factors that affect the cellular metabolism and the slow decline ('wash-out') of calcium currents during intracellular perfusion (see Chapter 5); it seems to be related to some direct action of Ba^{2+} ions as charge carriers on the calcium channel. The presence of a more resistant component in the total inward current may indicate that the toxic effect of Ba^{2+} could be expressed to a greater extent in a certain type of calcium channel (see next chapter).

Chemically operated calcium conductance in the neuronal membrane

Besides voltage-operated calcium channels, Ca^{2+} ions can also enter the nerve cell through the channels opened by the action of certain chemical substances on special receptor groups. In many neurons of the central nervous system these are glutamate-operated channels specifically sensitive to one of its synthetic agonists *N*-methyl-D-aspartate (NMDA). The detection of these channels has been delayed due to the fact that trypsin, which is often used for the isolation of central neurons, irreversibly inactivates them (Allen *et al.* 1988). The corresponding currents have been studied by MacDermott *et al.* (1986), Asher and Nowak (1987), and others. They are much less selective than the currents through voltage-operated channels: P_{Ca}/P_{Na} is about 10:6 (Westbrook and Mayer 1988); for comparison, this relation for the well-known acetylcholine-operated channels is only 0.22:1. It is still unknown what principles should be used for the description of the 'selectivity filter' of these channels. An interaction of the penetrating ion with the channel certainly takes place in this case; it can be predicted from the finding that Mg^{2+} and some other divalent ions block NMDA-channels in a potential-dependent way (Ascher and Nowak 1988). Further analysis has shown the complicated nature of such a block: the effect of different cations decreased in the sequence $Zn^{2+} > La^{3+} = Cd^{2+} > Mn^{2+} > Co^{2+} > Ni^{2+} = Mg^{2+}$; furthermore, the sites of action of inorganic and organic blockers seem to interact with each other (Greenberg and Marks 1988). NMDA-channels also pass monovalent cations quite well and they do not undergo any modification in the absence of external divalent ions. Examples of NMDA-induced calcium currents recorded by Kiskin in our institute in an isolated rat hippocampal neuron are presented in Fig. 2.9.

Relatively selective calcium currents can also flow through some other chemically operated channels. For instance, in smooth muscle cells ATP-sensitive channels have been described with selectivity $P_{Ca}/P_{Na} = 3:1$ (Benham and Tsien 1987).

Conclusions

The generation of a propagated action potential in response to membrane depolarization is an event in most neurons that opens an effective pathway for the entry of calcium ions into the cell along their steep electrochemical gradient. This pathway is formed by highly selective voltage-operated channels, which also pass Ba^{2+} and Sr^{2+} but do not pass monovalent ions. The

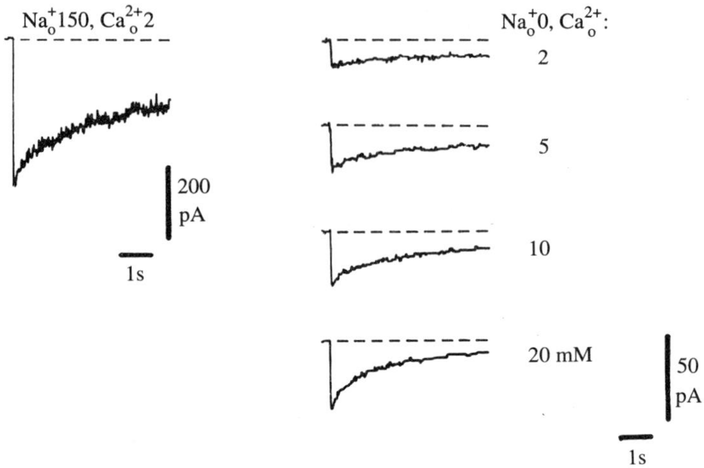

Fig. 2.9. Sodium and calcium currents through NMDA-operated channels in a neuron isolated from rat hippocampus. Concentrations of Ca^{2+} and Na^+ in the external solution are indicated near the records.

penetrating ions bind to specific groups in the channel which may be carboxyl residues; such binding is a necessary step for ion transfer, but it should not be too strong otherwise the ion would block the channel. Such a strong binding is the reason for the channel block by Co^{2+}, Mn^{2+}, Cd^{2+}, Ni^{2+}, and some other di- and tri-valent cations. Another binding site exists, somewhere in the calcium channel, that binds divalent ions especially effectively (like calcium chelators or calcium-binding proteins). The presence of a bound ion at this site is important for the maintenance of the selectivity of the calcium channel: elimination of a divalent ion from this site results in the transformation of the calcium channel into a sodium-selective one.

Entry of calcium ions in the neuron from the external medium can also occur through receptor-operated channels although in this case the inward current is much less selective and contains a large sodium component. The most studied channels of this type belong to a subgroup of glutamate-operated channels (NMDA-channels).

3. Gating mechanisms of calcium channels

History

As is clear from the preceding chapter, a distinctive feature of calcium channels in excitable membranes is their ability to activate (to open) upon a change of the transmembrane electric field. These channels are closed at high membrane potential level and pass Ca^{2+} ions across the membrane only during depolarization. Since the study of calcium channels started later than that of other voltage-operated channels (sodium, potassium), the understanding of the mechanism of their activation was much influenced by the data obtained earlier with respect to the latter channels. For sodium and potassium channels it is generally accepted that, together with structural components which determine the ability to select and to pass a definite ion type (the so-called 'selectivity filter'), there are structural components able to respond to a change of intramembrane electric field by opening or closing a pathway for the movement of ions along their electrochemical gradient ('gating' mechanism). This idea arose from the studies of Hodgkin and Huxley (1952), who suggested a model for such a gating mechanism which is now widely accepted. According to this model, ions can pass through the membrane only if a certain number of charged 'activating' particles (*m*-particles for sodium conductance, *n*-particles for potassium conductance) move simultaneously to a definite position in the channel under the action of the external electrical field. Besides, it is also necessary that a peculiar inactivating particle (*h*-particle), which blocks ion transport, be withdrawn. The equation for sodium conductance is described as:

$$G_{\mathrm{Na}} = \bar{G}_{\mathrm{Na}} m^3 h \tag{3.1}$$

where $\bar{G}_{\mathrm{Na}}$ is the maximal sodium conductance of the membrane, and m and h are activation and inactivation variables, respectively. Kinetic equations for the variables can be written as:

$$dm/dt = \alpha_m(1 - m) - \beta_m m \tag{3.2}$$

$$dh/dt = \alpha_h(1 - h) - \beta_h h \tag{3.3}$$

where α_m, β_m, α_h, and β_h are the rate constants for the corresponding reactions.

On the basis of the Hodgkin–Huxley model it was possible to reproduce with high accuracy a nerve impulse generated in the squid giant axon membrane, assuming the existence of only two independent types of ionic conductances (sodium and potassium). This model has been used successfully for the description of electrical phenomena occurring in many other excitable membranes. The authors themselves suggested that an independent proof of the reality of the physical mechanism underlying it can be obtained in the experiment by recording the intramembrane displacement of putative charged 'gating' particles. In fact, Armstrong and Bezanilla (1973) and Keynes and Rojas (1974) observed weak currents in the membrane of a giant axon, the characteristics of which corresponded to 'gating' currents. The charge carried by these currents at *on*-depolarization was equal to that carried in an opposite direction during repolarization (at least for short depolarizing shifts). The number of charges carried was increased with an increase of depolarization, attaining saturation at about the same potentials at which the sodium conductance became maximal. The question of how the movement of the putative inactivating particles is reflected in gating currents turned out to be less clear. It was not possible to record a proper 'inactivating gating current'. However, at long depolarizing shifts of the membrane potential, when a large inactivation of ionic current developed, the value of transferred gating charge turned out to be less at *off*-depolarization than that at *on*-depolarization, as a proportion of the gating particles become immobilized. These and other difficulties were the reasons for the construction of more and more complex models that suggested the existence of several consecutive states of ion channel, with possible direct and reverse transitions between them. The experimental data listed below show how these initial principles have been reflected in studies of presumed gating mechanisms of calcium channels.

Activation kinetics of calcium currents

The analysis of the kinetic characteristics of calcium currents is complicated by the fact that not one but several subtypes of ion channels are involved in the creation of the total calcium current through the surface membrane. The first data showing that, according to kinetic characteristics, calcium currents can be separated at least into two components were obtained in egg cells of Polychaeta (Fox and Krasne 1981) and rat dorsal root ganglion neurons in our laboratory (Veselovsky and Fedulova 1983; Fedulova *et al.* 1985), and then independently in the laboratories of Lux (Carbone and Lux 1984, 1987*a*, *b*) and Tsien (Nowycky *et al.* 1985*a*). These results were then corroborated for the membranes of different excitable cells: freshly isolated (Mironov *et al.* 1985) and cultured (Lozinskaya and Shuba 1989) snail neurons, cultured rat sensory neurons (Bossu *et al.* 1985), neurons of some brain structures (Akaike *et al.* 1989), retinal bipolar cells (Kaneko *et al.* 1989), cardiomyocytes (Bean 1985; Mitra and Morad 1986; Bonvallet

1987), smooth muscle fibres (Loirand *et al.* 1986), skeletal muscle fibres (Cognard *et al.* 1986; Cota and Stefani 1986; Cavalie *et al.* 1987), clonal pituitary cells (Matteson and Armstrong 1986; Cohen and McCarthy 1987; Suarez-Kurtz *et al.* 1987), pars intermedia cells of the pituitary gland (Cota 1986), adrenal glomerulosa cells (Cohen *et al.* 1988), and pancreatic α and β cells (Hiriart and Matteson 1988; Rorsman 1988). Two components of calcium currents have even been distinguished in such unexcitable cells as fibroblasts (Chen *et al.* 1988; Peres *et al.* 1988); they were also discovered in unicellular organisms (Deitmer 1986).

The most significant distinction between the observed calcium current components is their activation at different membrane potential levels, allowing them to be defined as 'low-threshold' and 'high-threshold' currents respectively. The former currents are activated in the range between −60 and −40 mV, i.e. in the region close to resting potential level. The latter currents, on the contrary, require a stronger polarization of the membrane for their development (about up to 0 mV).

Figure 3.1 presents current–voltage characteristics of inward currents recorded from a rat dorsal root ganglion neuron at various concentrations of calcium ions in the extracellular medium; a bend in the curves clearly shows the area of activation of the high-threshold component. The rate of activation for the low-threshold current was about two times higher than that for the high-threshold one. Therefore, it was originally suggested that low-threshold currents could be defined as 'fast' currents, in contrast to 'slow' high-threshold ones; however, such a classification turned out to be incorrect

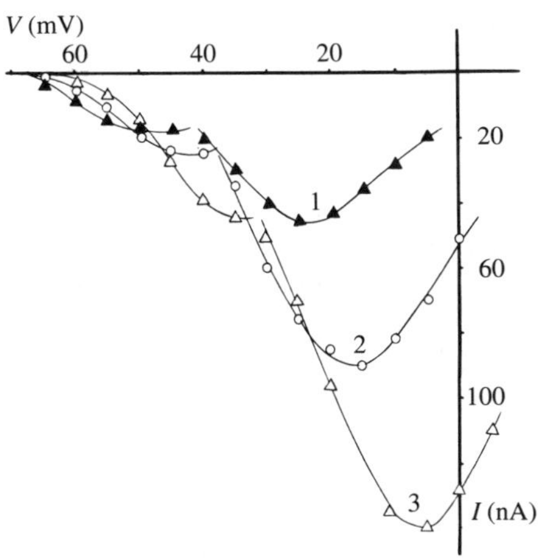

Fig. 3.1. Current–voltage characteristics of calcium currents in rat dorsal root ganglion neuron recorded at a holding potential −90 mV (Veselovsky and Fedulova 1983). $[Ca^{2+}]_{out}$: 2 mM (1). 5.4 mM (2), 14.8 mM (3). Intracellular perfusion with 150 mM Tris-PO_4.

because according to their most important kinetic characteristics (mean open time) it is in fact the low-threshold channels that are 'slower' compared with the high-threshold channels (see details below).

Later, the high-threshold component of the calcium current was also found to display heterogeneity with respect to the activation threshold and, in particular, to its behaviour during a prolonged shift of the membrane potential: one portion of the current is subjected to rather fast inactivation (for details see Chapter 4), the other one inactivates very slowly and therefore can create a steady entry of ions into the cell (Nowycky *et al.* 1985*a*; Fox *et al.* 1987*a*, *b*; Kostyuk *et al.* 1988*a*).

These components of calcium currents turned out to be expressed in different ways in various cells. Thus, in the membrane of clonal neuroblastoma cells most currents that have been recorded are similar to the low-threshold current component in mouse sensory neurons (Moolenaar and Spector 1978; Spector 1981). According to our data, non-differentiated neuroblastoma cells N1E-115 showed only low-threshold currents (Veselovsky *et al.* 1984; Kostyuk *et al.* 1988*b*). In clonal PC12 phaeochromocytoma cells, only a high-threshold, non-inactivating current component has been recorded (Shuba and Lyubanova 1988). However, in experiments on other clones of blastomatose cells (hybrid of neuroblastoma and glioma) with the use of quinidine-induced suppression of outward currents, as well as a low-threshold current a slowly inactivating high-threshold current was observed, like that described in sensory neurons (Fishman and Spector 1981). It is interesting that some authors also separated two components of calcium current in N1E-115 neuroblastoma cells (Tsunoo *et al.* 1984; Yoshii *et al.* 1988); it is possible that such a discrepancy with our data might be related to the use of more differentiated cells in which the composition of ion channels in the membrane can be quite different from that in non-differentiated cells (see details in Chapter 7).

This circumstance can be used for an easier analysis of kinetic characteristics of calcium currents, since it allows one to choose certain cell lines as models for the selective separation in 'pure' form of current components created by a certain channel type. Another effective approach is the separation of currents by choosing a potential level held on the membrane of a studied cell. Low-threshold currents are characterized by a fast potential-dependent inactivation which already develops at membrane potential levels close to the resting potential and which necessitates a considerable hyperpolarization of the membrane for their functioning. Due to this fact the low-threshold component can be virtually completely suppressed by a preliminary decrease of the membrane potential (about up to -50 mV); in this case the high-threshold current is fully preserved. As a low holding-potential level was used initially for measurements of the activation kinetics of calcium currents, these currents were not distorted by the possible involvement of low-threshold channels in their generation.

A characteristic feature of high-threshold calcium currents in all investigated excitable membranes is their slow rise (in contrast to 'fast' sodium currents) at corresponding depolarization shifts of the membrane potential. Therefore, these currents are still sometimes (for example, in cardiomyocytes) designated as 'slow inward currents', I_{si}. Quantitative analysis of the activation kinetics of these currents has been performed on a number of cell types, with the use of approaches elaborated by Hodgkin and Huxley.

For the somatic membrane of snail neurons it was shown in our studies that the activation kinetics of the high-threshold calcium current can be satisfactorily described by this model, using not a cubic but a square power of the m-variable:

$$I_{Ca} = \bar{G}_{Ca} m^2 h (E - E_{Ca}) \tag{3.4}$$

The accuracy of approximation in this case proved to be very high (Shakhovalov 1977; Kostyuk *et al.* 1979). This conclusion can be seen from the currents presented in Fig. 3.2. The inactivation variable, h, could be neglected due to the slow velocity of this process in calcium channels (see below). In rat dorsal root neurons the most adequate power of the activation variable also turned out to be 2 (Kostyuk *et al.* 1981*e*). The square power of

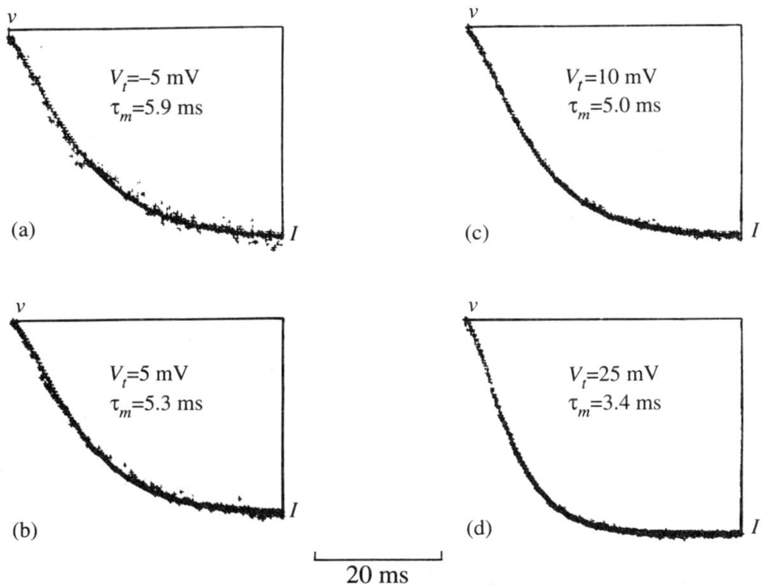

Fig. 3.2. Approximation of the activation kinetics of the high-threshold calcium current in the snail neuron by the Hodgkin–Huxley model (Kostyuk *et al.* 1979). Measurements were made at 6.5 °C, with a holding potential of −40 mV. Experimental data are presented by crosses, the theoretical curves by continuous lines, for four different conditions (a–d).

the *m*-variable is most suitable for the description of activation kinetics for high-threshold calcium currents in other molluscan neurons (Byerly and Hagiwara 1982; Byerly *et al.* 1984), guinea-pig pyramidal neurons (Kay and Wong 1987), and rat neurohypophyseal cells (Hagiwara and Ohmori 1982). However, Akaike *et al.* (1978) obtained a less satisfactory approximation of the activation kinetics of calcium current in snail neurons. In non-dialysed *Aplysia* neurons the power of the activation variable for different test potentials ranged between 1 and 2 (Adams and Gage 1979*c*). In the presynaptic membrane of the squid giant synapse (Llinas *et al.* 1976), as well as in the membrane of crayfish muscle fibres (Henček and Zachar 1977), a satisfactory description of activation kinetics was possible by using higher powers of the *m*-variable. For calcium currents of frog skeletal muscle fibres, the cubic power of the *m*-variable was adequate (Sanchez and Stefani 1983).

Activation kinetics of the low-threshold current component in rat sensory neurons could also be approximated by the Hodgkin–Huxley equation, using the square power of the *m*-variable; however, the activation time constant itself was about half as long as that of the main component (Veselovsky and Fedulova 1983). The same has been found for N1E-115 neuroblastoma cells (Yoshii *et al.* 1988).

Deactivation kinetics of calcium channels

Kinetic characteristics of the gating mechanism of calcium channels can also be assessed by the rate of disappearance of the ionic current after repolarization of the membrane to the level at which ion channels become closed (by the so-called current 'tails'). Proceeding from the principles of the Hodgkin–Huxley model of gating particles, closure of the channel must occur if at least one particle is transferred from the 'activated' to the 'resting' state; therefore the time constant of the current 'tail' should be less than its activation time constant. On the assumption that two activating particles exist in the calcium channel, the former time constant should be about half as long as the latter constant.

In several laboratories, special techniques have been used that allowed one to minimize the hindrance created by capacitance currents and so to measure calcium current 'tails' at the shortest time intervals after the onset of membrane repolarization. Under these conditions in *Limnea* neurons an exponential component of current decay was observed at a reduced temperature (7–10 °C) with a time constant of 0.2–0.4 ms followed by a slower component with a time constant of 2–4 ms (Byerly and Hagiwara 1982; Byerly *et al.* 1984). In *Aplysia* neurons under the same conditions a fast component of calcium current decay with a time constant 0.38 ± 0.5 ms and a slower component with a time constant of 2.0 ± 0.8 ms (12.5–14 °C) have been distinguished. A still slower component with a time constant of about 30 ms, which was occasionally observed, apparently had nothing to do with the calcium current and could be evoked by a non-specific current (see Chapter 4 of this volume; Eckert and Ewald 1983*a*). Two exponential components in current 'tails' have also been revealed in perfused neurons of the snail *Helix aspersa* (Brown *et al.* 1983).

The data obtained are very complicated, therefore in our laboratory we analysed in detail the calcium current 'tails' of rat dorsal root ganglia neurons in which, as was indicated above, three calcium current components can be distinguished, differing by their potential-dependence, activation and inactivation kinetics (Veselovsky *et al.* 1985; Kostyuk and Shirokov 1989). In order to minimize the hindrances created by capacitance currents, we used summation of responses to depolarizing and hyperpolarizing potential shifts. Since complete activation of all types of calcium channels in these neurons necessitates the maintenance of a holding potential at about −100 mV, and considerable hyperpolarization up to this level can damage the membrane, we used the procedure of 'divided' displacements: the response to one depolarization was summed with four responses to a fourfold less hyperpolarizing displacement. In such neurons (differing from molluscan neurons by their considerably smaller size and correspondingly by a lower input capacitance of the membrane) this procedure turned out to be sufficient for reliable separation of the fastest components of the current decay; the capacitance artefact did not exceed 200–300 μs.

Separation of 'tails' related to different components of the calcium current was achieved by selecting the level of holding and test potentials: at a holding potential of about −90 mV and depolarization up to −40 to −30 mV practically only the low-threshold current was activated; at lower holding potentials this current was subjected to stationary inactivation, and then at a more considerable depolarizing shift the main high-threshold components of calcium current could be evoked in pure form.

Deactivation of the low-threshold current was purely mono-exponential by its kinetics. This feature was found at any holding potential and depolarizations, and did not depend on the composition of the perfusion solution. The amplitude of the current 'tail' was proportional to the current value at the end of the depolarizing shift, and its time constant was characteristically changed with a change of repolarization potential: it decreased as the potential was shifted to more negative values and then became constant (1.1 ± 0.1 ms) at very negative membrane potentials (−140 to −160 mV). The time constant proved to be dependent on the kind of penetrating ion: it increased by about twofold upon replacement of Ca^{2+} ions by Ba^{2+} ions. Some examples are shown in Fig. 3.3.

Deactivation of high-threshold currents was measured at a decrease of holding potential up to −80 to −70 mV. Such a decrease did not significantly affect the amplitude of the high-threshold current, but at the same time it steadily inactivated the low-threshold current by 70–80 per cent; therefore

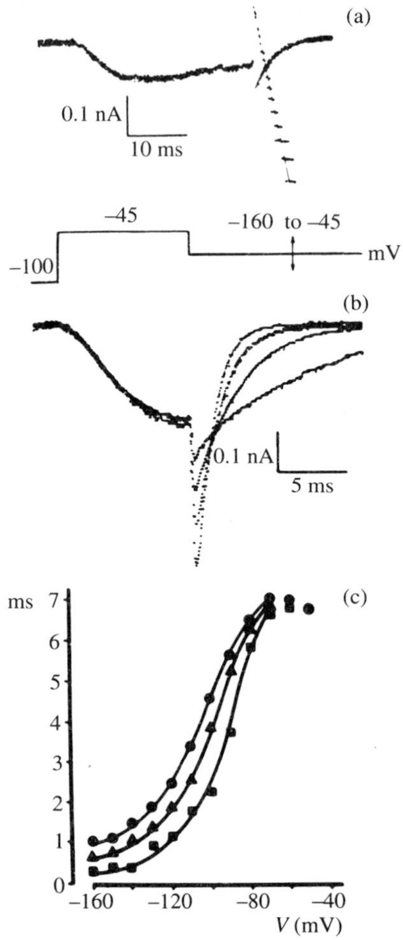

Fig. 3.3. Potential-dependence of deactivation kinetics of low-threshold calcium currents in mice dorsal root ganglion neurons (Kostyuk and Shirokov 1989). (a) Example of low-threshold inactivating (LTI) current together with semilogarithmic plotting of the tail current. (b) changes in tail-current kinetics at different repolarization potentials (the pulse protocol is presented above the figure). (c) Potential dependence of deactivation time constant measured with different permeant ions. Ca^{2+} (●), Sr^{2+} (▲), Ba^{2+} (■).

the latter could be virtually ignored. Deactivation kinetics of the high-threshold current could be satisfactorily approximated by only two exponents. A fast component had the time constant of 0.13 ± 0.05 ms ($n = 18$) and a slow one of 0.83 ± 0.04 ms ($n = 11$). Partial separation of these components could be performed by a more marked decrease of the holding potential (up to −40 mV). In this case the slow component was essentially reduced or completely removed, while the fast one was practically unaffected. Both components also depended on the value of activating depolarizing displacement in a different way. If the maximal amplitude of the fast component increased up to a certain limit and then did not change as depolarization increased, the changes in the maximal amplitude of the slow component had a bell-shaped character—upon an increase in depolarization it decreased again. These phenomena are shown in Fig. 3.4.

These peculiarities can probably be explained on the basis of the results mentioned above concerning the presence of two components of the high-threshold calcium current in the membrane of the cells studied. One of these is subjected to inactivation (it deactivates more slowly and its amplitude decreases with an increase in depolarization) and the second one (with the fastest deactivation) has a stationary character.

An interesting finding was a different action of the kind of *anion* introduced into the cell during perfusion on the two components of high-threshold current deactivation. Replacement of Tris-aspartate in the intracellular solution with Tris-phosphate reversibly removed the slow component of the current 'tail' while not changing its fast component. At the same time, the time course of the high-threshold current was also changed; it decreased at the expense of its inactivating component. This can be seen from the current–voltage characteristics of the calcium current obtained in the same cell during replacement of perfusing solution, as shown in Fig. 3.5. The reason for such an effect is not clear. One could think that it related to the change in the level of free Ca^{2+} ions inside the cell when solutions of varying anionic composition were used; however, this effect was preserved even after stabilization of this level with the help of an EGTA buffer.

The discovery of the method for separation of two components in the 'tail' of the high-threshold calcium current made possible a detailed investigation of their individual potential-dependence. As in the case of the low-threshold current, their time constants reduced as the repolarization potential was shifted to more negative values and became finally constant. The time constant of the slow component became constant at potentials of about −70 mV and that of the fast component at −50 mV. Upon replacement of the kind of penetrating cation no appreciable changes in the slow component of the 'tail' current have been found; it was impossible to measure changes in the maximal amplitude of the fast component because of its very fast time course. Examples of potential-dependence of these currents are demonstrated in Fig. 3.6.

The results are in conformity with the suggestion that the kinetics of functioning of calcium channels of different types are based on similar mechanisms: only the time constants of transitions between different functional states differ in different channels. In particular, the transition of channels from the open to the closed state in all cases obeys first-order kinetics; the tail currents with more complex kinetics indicate a simultaneous functioning of different types of channels. In these transitions there is a component which, unlike the macroscopic kinetics of calcium-current activation, is little dependent on the potential; it is well expressed in the region of high negative values of the membrane potential. Further develop-

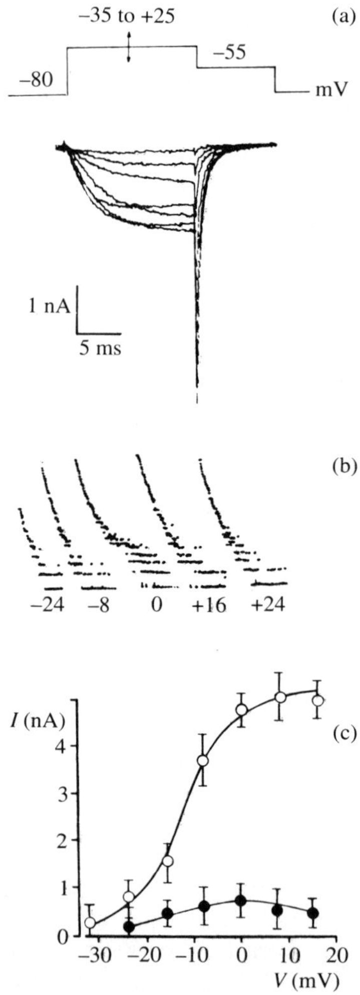

Fig. 3.4. Separation of exponential components in the deactivation of high-threshold calcium current (Kostyuk and Shirokov 1989). (a) Examples of high-threshold currents recorded at different activating potential shifts (see the protocol above the figure). (b) Tail currents on a semilogarithmic scale; the corresponding depolarization levels are indicated below the records. (c) Amplitudes of the fast (○) and slow (●) exponential components obtained by regression analysis.

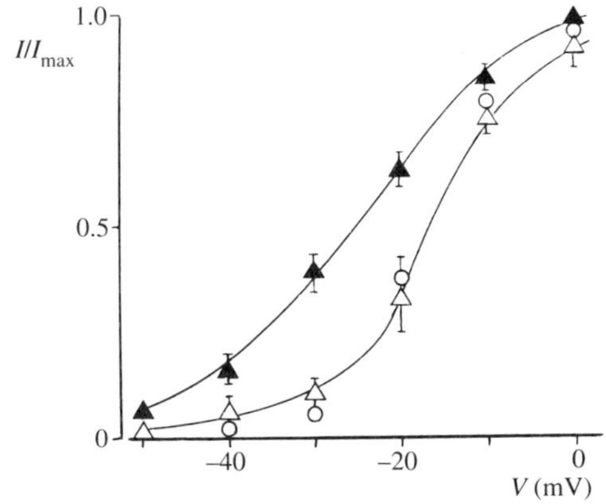

Fig. 3.5. Effect of different intracellular anions on high-threshold calcium currents (Kostyuk and Shirokov 1989). Activation curves obtained from six cells with Tris-aspartate (▲) and Tris-phosphate (△) solutions containing EGTA, as well as Tris-phosphate without EGTA (○).

ment of these ideas took place after the analysis of the kinetics of functioning of various kinds of single channels (see below).

Gating currents of calcium channels

The successful mathematical description of activation and deactivation kinetics of the calcium current does not indicate the direct mechanism of channel transition from non-conducting to conducting state. In this respect it is especially important to compare such kinetics with the characteristics of asymmetrical intramembrane charge displacement ('gating currents') which, as was mentioned above, can accompany the opening and closure of ion channels.

When studying electrical phenomena that develop upon depolarization of the neuronal membrane, it was found that, in parallel with activation of an ionic conductance, an intramembrane charge displacement occurs which has a complex time course. Such a charge displacement can be observed if the ionic current itself is suppressed by any blocker. This displacement has a slow component which, according to its characteristics, corresponds to a 'gating current' of calcium channels (Kostyuk *et al.* 1977; Adams and Gage 1979*b*). It was later shown by Kostyuk *et al.* (1980*b*, 1981*b*) that charge displacement in a snail neuronal membrane during its depolarization can be separated into fast and slow components differing in their potential-dependence. Perfusion of the cell with fluoride-containing solution, which

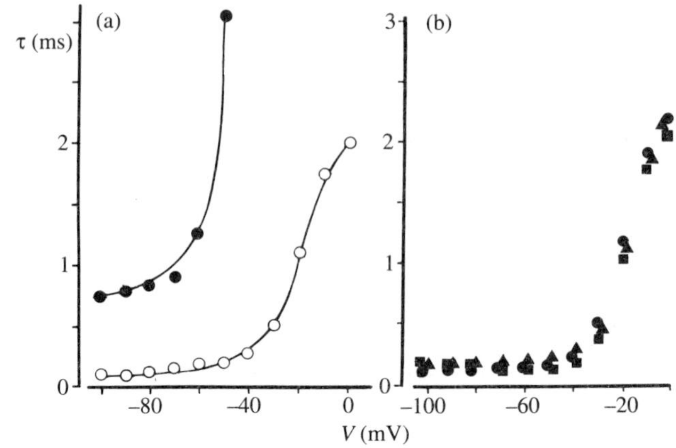

Fig. 3.6. Time constants of two exponential components in the deactivation kinetics of high-threshold calcium currents. (a) Potential-dependence of the time constant of slow (●) and fast (○) components recorded at holding potentials −80 and −40 mV, respectively. (b) Potential-dependence of the fast time constant with Ca^{2+} (●), Sr^{2+} (▲), and Ba^{2+} (■) as permeant ions.

irreversibly suppresses calcium conductance, led to a parallel depression of the slow component of the asymmetrical displacement current; only the fast component was preserved, which is analogous to the 'gating current' of sodium channels by its time course. Examples of the corresponding currents are given in Fig. 3.7. If, from the total displacement current, a component was subtracted which was preserved during cell perfusion with fluoride, the slow component thus obtained in pure form corresponded well to the 'gating current' of high-threshold calcium channels in its characteristics. This correspondence included the following:

1. The charge value displaced during the development of slow component was equal for both *on*- and *off*-depolarizations (see Fig. 3.8); thus, it represented just the intramembrane charge displacement.
2. If depolarization increased, the value of the displaced charge also increased, attaining saturation at the same potentials at which the calcium current saturated (Fig. 3.9); the maximal value of displaced charge was about 1500–2000 electron charges per μm^2 membrane unit.
3. A graph for potential-dependence of the quantity of transferred charges at *on*-depolarization completely coincided with that for potential-dependence of the *m*-variable [calculated as the square root of normalized calcium conductance according to eqn 3.4; (Fig. 3.9)]
4. The time constant of displacement current decay, τ_{on}^{ass}, at a wide range of membrane potential changes fitted well to the activation time constant of

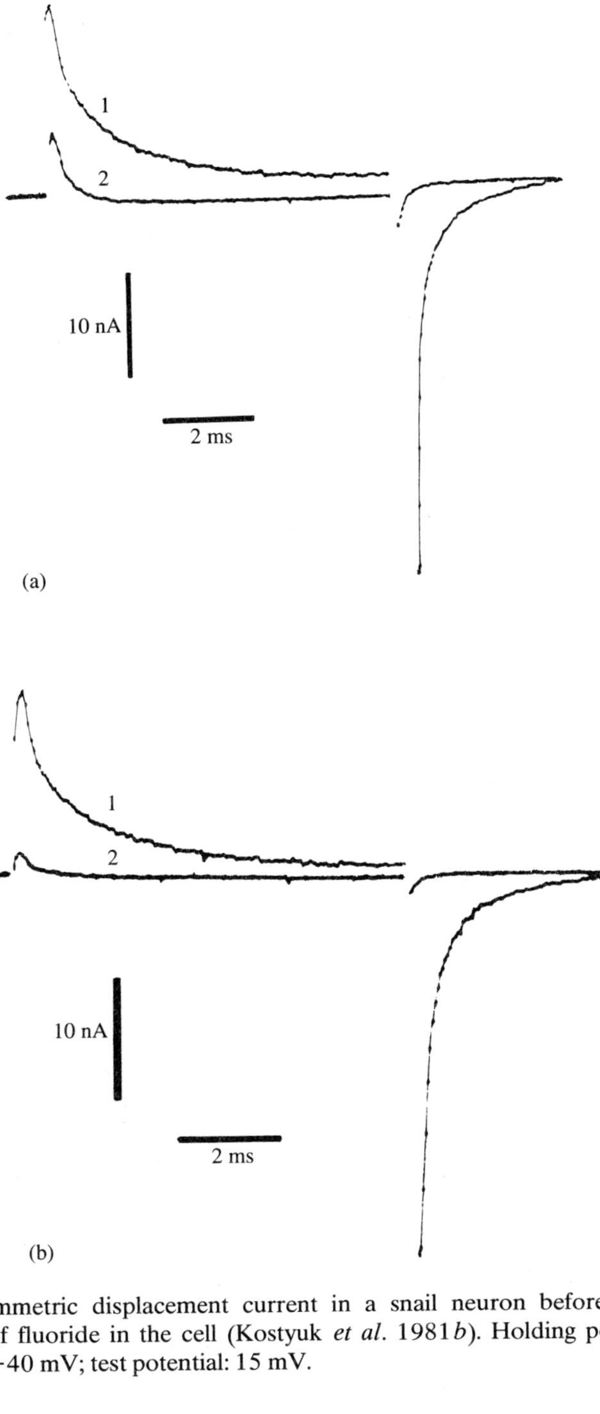

Fig. 3.7. Asymmetric displacement current in a snail neuron before and after introduction of fluoride in the cell (Kostyuk *et al.* 1981*b*). Holding potential: (a) −90 mV, (b) −40 mV; test potential: 15 mV.

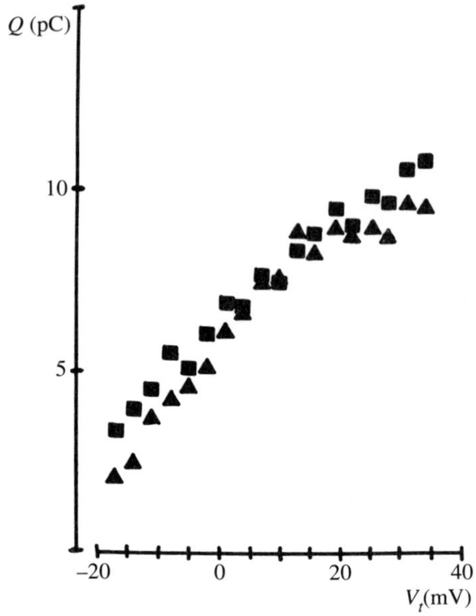

Fig. 3.8. Voltage distribution of fluoride-sensitive charges displaced by the *on*- (▲) and *off*- (■) components of the asymmetric displacement current (Kostyuk *et al.* 1981*b*). Holding potential: −40 mV.

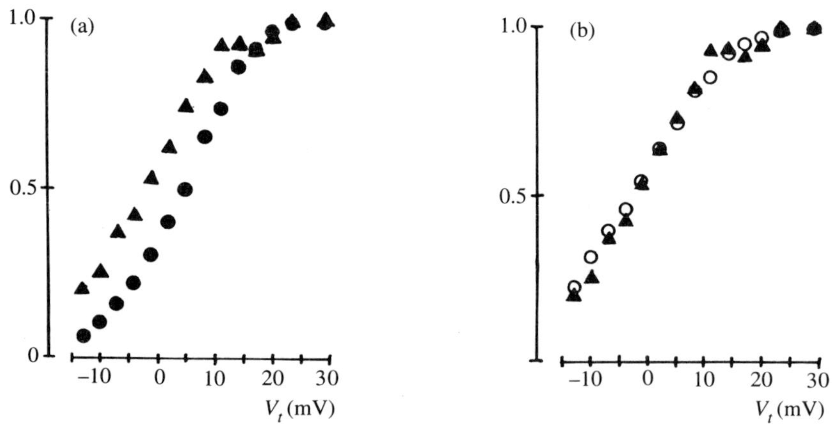

Fig. 3.9. Voltage distribution of fluoride-sensitive asymmetric charge displacement (triangles) and conductance (circles), normalized (Kostyuk *et al.* 1981*b*). In (b) the square root of conductance is presented, contrary to (a).

calcium conductance, τ_m. The time constant of displacement current decay at *off*-depolarization, $\tau_{\text{off}}^{\text{ass}}$, was about half as long as that of the calcium current relaxation. Figure 3.10 presents the kinetic characteristics of calcium currents and the corresponding displacement currents.

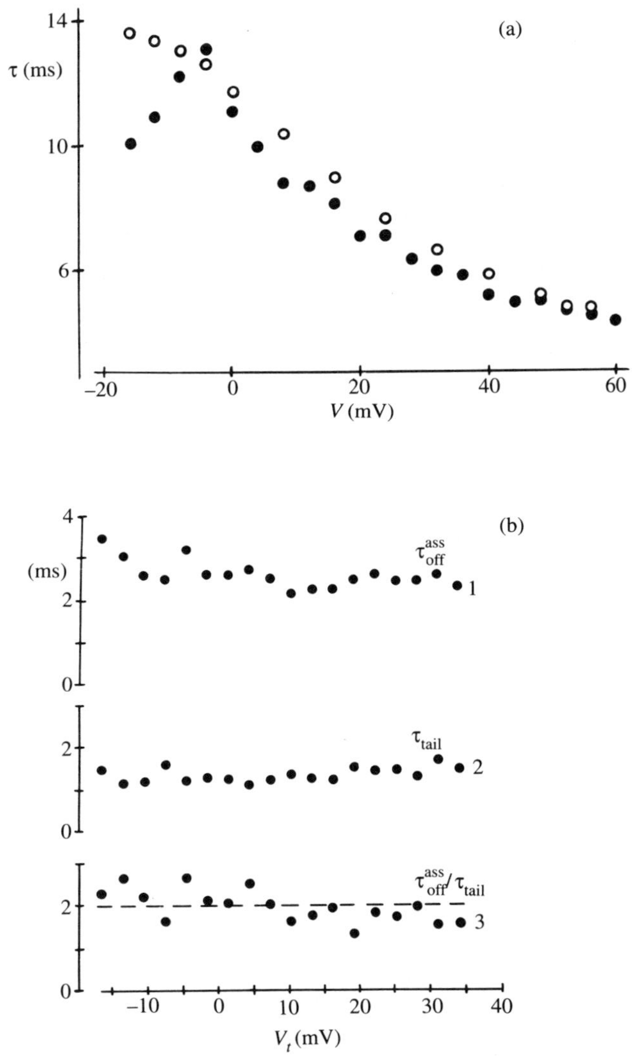

Fig. 3.10. Voltage dependence of the time constants of asymmetric charge displacement and calcium current (Kostyuk *et al.* 1981*b*). (a) Dependence of τ_m (○) and $\tau_{\text{on}}^{\text{ass}}$ (●). (b) Dependence of $\tau_{\text{off}}^{\text{ass}}$ (1), $\tau_{\text{off}}^{\text{Ca}}$ (2), and $\tau_{\text{off}}^{\text{ass}}/\tau_{\text{off}}^{\text{Ca}}$ (3). Holding potential: −40 mV.

The results obtained fit exceptionally well to predictions of the modified Hodgkin–Huxley model, and may have a certain physical meaning. Determination of maximal logarithmic slopes of the functions $Q(V)$ and $G_{Ca}(V)$ gave the corresponding values 8 mV and 4 mV to an *e*-fold change of the displaced charge value and calcium conductance, respectively (see Fig. 3.11). If such a displacement is assumed to obey the Boltzmann distribution, and to represent redistribution of some charged groups in calcium channels between two extreme stationary states, one can determine the effective valency for charged groups: it proves to be equal to three.

Our findings are in good agreement with the idea that simultaneous displacement of two 'gating' particles, each having the effective valency of about three, is necessary for transition of the calcium channel into an open state. If even one of these particles is in the initial state, the channel becomes closed. Thus, in its main features, the gating mechanism of calcium channels seems to be analogous to that of sodium channels studied earlier, and differs from it only quantitatively (by the number of presumed gating particles).

A similar investigation of gating currents in calcium channels has been

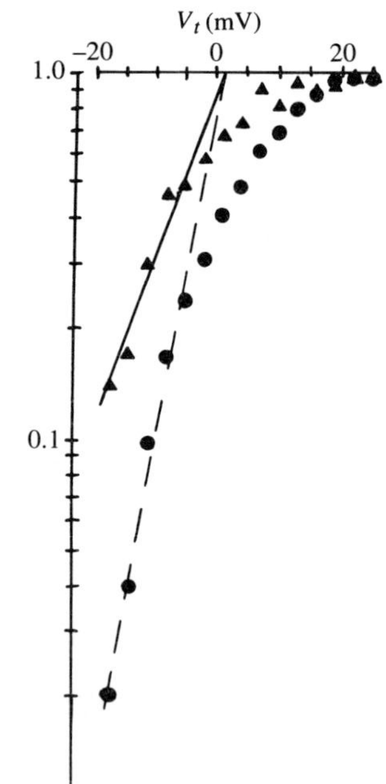

Fig. 3.11. Voltage distribution of the asymmetric charge displacement (solid triangles) and calcium conductance (solid circles) presented on a semilogarithmic plot, normalized (Kostyuk *et al.* 1981*b*). The slopes are 8 mV and 4 mV per *e*-fold changes.

performed recently on cardiomyocytes of new-born rats. In this case the analysis was difficult due to the presence of well-expressed low- and high-threshold components of the calcium current, as well as to the absence of an effective way to selectively block the calcium current (dihydropyridine blockers of calcium channels were used, which do not suppress the low-threshold component and which induce an incomplete block of the high-threshold component). Nevertheless, the results showed that a considerable portion of the gating current in the membrane of these cells is related to dihydropyridine receptors, i.e. with calcium channels. As in the case of molluscan neurons, the value of the charge carried was very large compared with the value of the calcium current; the most likely explanation for such a discrepancy is the existence of 'closed', non-functioning calcium channels, in which the transfer of gating charges in response to depolarization of the membrane still occurs (Field *et al.* 1988).

The study of gating currents is naturally only the first approach to the understanding of the mechanism of calcium current activation. The more complex nature of this mechanism can be predicted by some peculiarities of the gating currents. Accordingly the good quantitative coincidence described above is true for substantial displacement of the membrane potential. When these processes were investigated at very small displacements, considerable deviations from this coincidence occurred; in particular, relaxation time constants for these currents turned out to be little dependent on the membrane potential under such conditions (Kostyuk *et al.* 1981*b*).

Investigation of calcium channel behaviour by current fluctuations analysis

Information about the behaviour of unitary ionic channels can be obtained by the analysis of current noises in the membrane. Such noises reflect the discrete nature of molecular structures creating transmembrane ionic currents; they can be revealed when there is a substantial decrease in the number of structures participating in the generation of the total current. The efficacy of membrane noise analysis is mainly determined by the amplitude of the current produced by a single channel. At sufficient amplitude, current fluctuations related to the opening of single channels will prevail over other types of noise (temperature noise, noise from amplifiers, the so-called $1/f$-noises). This is the reason why the most effective analysis was that of noises created by activation of chemically operated channels. Thus, conductance of the cholinoreceptor ion channel in muscle fibre end-plates is equal to several dozens of pico-siemens; therefore, current fluctuations evoked by switching on or off of such single channels could be recorded rather easily. The analysis of noises created by voltage-operated sodium channels showed that, in this case, the unitary channel conductance is an order lower and, correspondingly, the requirements for recording and analysis of such noises are more rigorous (see Sigworth 1977).

For statistical analysis of the noise created by the activity of high-threshold calcium channels in the somatic membrane of snail neurons, we used in our laboratory the recording of noises from small membrane patches sucked into a micropore of a special plastic pipette (Krishtal *et al.* 1980, 1981). In order to increase the contribution of single-channel currents to the total current, the recording was made in conditions of maximal single-channel currents. For this purpose Ca^{2+} ions in the extracellular medium were replaced by Ba^{2+} ions (which, as was mentioned above, pass more readily through calcium channels). Furthermore, the latter ions were used at large (isotonic or hypertonic) concentrations. As a result, one could record current fluctuations of about 10^{-11} A from the recorded membrane patch, quite enough for statistical analysis.

Statistical analysis of recorded fluctuations has shown that their mean square deviation increases with the increase of membrane depolarization, reaching a maximum at membrane potentials when the calcium current is at half-maximum; upon further depolarization it decreased again (Fig. 3.12). Just this correlation should be expected if the recorded fluctuations actually reflect the binomial distribution of statistical openings and closures of single ion channels.

From the mean square deviation of the noise one can evaluate the current flowing through a single channel (if it is assumed to be constant):

$$i_{Ca} = \frac{\sigma^2}{\bar{I}(1 - I/I_{max})} \tag{3.5}$$

where $\bar{I}$ is the mean calcium current and I_{max} is its maximal value (when all channels are open).

With 130 mM Ba^{2+} in the extracellular solution, the calculated current value through a single calcium channel was 0.20 ± 0.02 pA (at 22 °C). If we assume that Ba^{2+} ions penetrate through the high-threshold calcium channel about twice as readily as Ca^{2+} ions, the calcium current through the channel at such a high concentration has to be only about 0.1 pA. The current value for a unitary calcium channel can be evaluated in another way. As the charge carried by the gating mechanism of the calcium channel is known (see above), one can determine the number of functioning calcium channels by knowing the general maximal value of the displacement charge. Then from the value of the calcium current through the membrane one can determine the current through a single channel, i_{Ca}:

$$i_{Ca} = \frac{2z'e\,I_{max}}{Q_{max}} \tag{3.6}$$

where I_{max} is the maximal calcium current, Q_{max} is the maximal value of displaced charge, e is the value of elementary charge, and z' is the effective valency of the gating particle. It should be taken into account that in all cases the intramembrane charge displacement is accompanied by alternating opening of the channels.

According to such calculations, the current through a unitary calcium channel should be 0.02 ± 0.01 pA and the density of calcium channels 250–330 μm^2 (Kostyuk *et al.* 1981*b*). As in this case the current was measured at the normal concentration of Ca^{2+} in the extracellular medium, the value thus obtained approximately corresponds to that evaluated by noise fluctuation analysis. Judging by these values a single calcium channel can pass at most 6×10^5 Ba^{2+} ions or 3×10^5 Ca^{2+} ions per second.

The analysis of fluctuations of ionic currents through calcium channels was also performed on other tissues, in particular on rat pituitary cells. In this case the calculated current through single channel at 100 mM Ba^{2+} in the external solution was equal to 0.7 pA (Hagiwara and Ohmori 1982).

Measurement of current fluctuations can yield information about the kinetic characteristics of ion channels. For this purpose the measurement of fluctuation frequency spectra has been performed. A theoretical analysis of this question made by Stevens (1972) has shown that the fluctuation spectrum due to statistical opening or closing of elementary conductance units must be described by the Lorentz function:

$$S(f) = \frac{S(0)}{1 + (f/f_{1/2})^2} \tag{3.7}$$

where $S(0)$ is the spectral density at zero frequency and $f_{1/2}$ is the so-called cut-off frequency at which spectral density $S(f_{1/2}) = 1/2S(0)$. From the latter value a presumed relaxation time constant of the channel can be determined

$$\tau_L = 1/2\pi f_{1/2} \tag{3.8}$$

The noise recorded by a computer was subjected to a fast Fourier transform and the density spectrum so obtained was approximated by the Lorentz function. Then the cut-off frequency found graphically was used to determine the relaxation time constant of the channel. In this case current fluctuations prevailed so much over other noises (for example, the so-called *l/f* noise) that the latter could be neglected. A crucial moment in such measurements was the necessity to record the noise for a long time in stationary conditions, i.e. at a constant value of ionic current itself. Meanwhile, ionic currents are not kept constant during stationary depolarization of the membrane but decline gradually (see Chapter 6). This difficulty could be avoided to some extent if one chooses quasistationary periods in the development of ionic currents which inactivate quickly at the beginning of depolarization and then inactivate more slowly. For the analysis, the periods

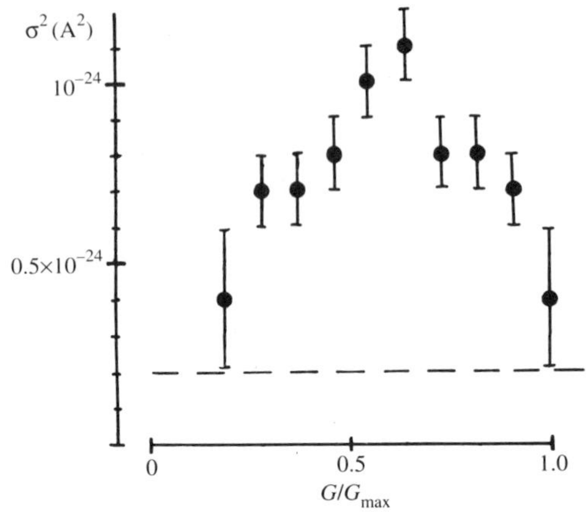

Fig. 3.12. Dependence of the dispersion of barium current fluctuations (σ^2_{Ba}) in a snail neuron on normalized conductance (Krishtal 1981).

of delayed inactivation were chosen when the current reached a quasi-stationary value. However, this approach entails other complications, which will be discussed below.

Examples of fluctuation spectra obtained in these experiments for barium currents through high-threshold calcium channels in the somatic membrane of snail neurons can be seen in Fig. 3.13. From these spectra the open time of a calcium channel was 0.7 ± 0.2 ms at 22 °C. The measurement of this characteristic at different polarizing shifts of membrane potential showed that the channel open time is little dependent on the potential level. This result is also not in conformity with a simple model of a gating mechanism of the calcium channel, according to which simultaneous displacement of two gating particles is needed to open the channel. It is evident that the process of channel activation is a multi-stage one. The displacement of charged gating groups is a necessary but not a sufficient stage in this process; there are also stages little dependent on the membrane potential.

As was mentioned above, calcium channels can be essentially modified by removal of Ca^{2+} ions from the extracellular medium with the use of chelating agents; they then start to pass sodium ions. It was interesting to find out how single-ion channels will change upon such modification. The difficulty was that during extracellular introduction of Na^+ ions, apart from the currents through modified calcium channels, those through natural sodium channels would appear. This could be avoided by selection of the cells studied, some of which did not show 'natural' sodium currents.

The studies of inward current fluctuations on these cells have shown that

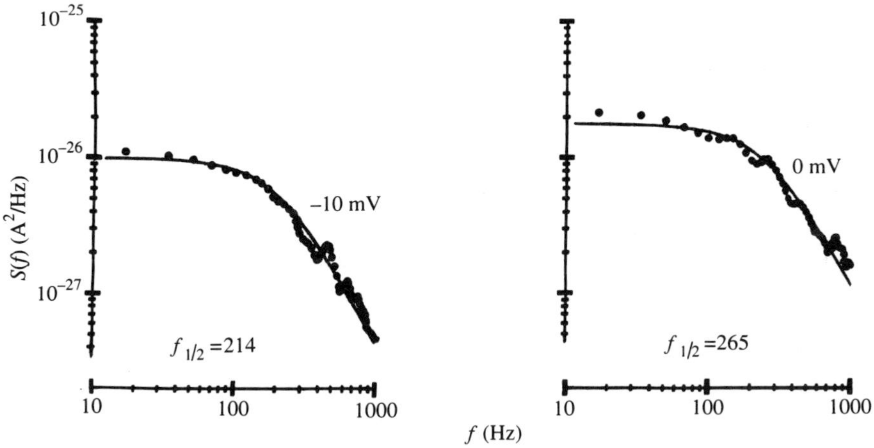

Fig. 3.13. Barium current fluctuation spectra in a snail neuron and their fits to Lorentz functions (Krishtal *et al.* 1981). Holding potential: −50 mV. Test potentials and cut-off frequencies are indicated near each spectrum.

sodium currents through single modified calcium channels are similar to those through natural sodium channels, i.e. much larger than calcium currents before modification. However, the open time of such channels determined by spectral density fluctuations only slightly increased, this value being 0.9 ± 0.2 ms at the same temperature (see examples of the spectra in Fig. 3.14). Obviously, such modification concerns more the selective properties of the channels than their gating mechanism (Kostyuk *et al.* 1981*b*).

Direct investigation of single calcium channel kinetics

The methods of measurement of membrane current fluctuations progressed with the measurement of current jumps created by opening or closing of single ionic channels. This measurement became possible due to elaboration of a special technique for voltage clamp on a membrane patch, called the 'patch-clamp' method.

The method consists of the use of special polished glass micropipettes, in which the inside of the opening is coated by Sylgard to ensure hydrophobic properties. The micropipette opening is 0.5–1.0 μm in diameter. The pipette is attached to a system that allows one to create negative hydrostatic pressure inside it, and thus to suck in a patch of the cell surface membrane. If the membrane patch is properly processed in order to remove folds and dirt (for such treatment different enzymatic methods are used), good adhesion can be

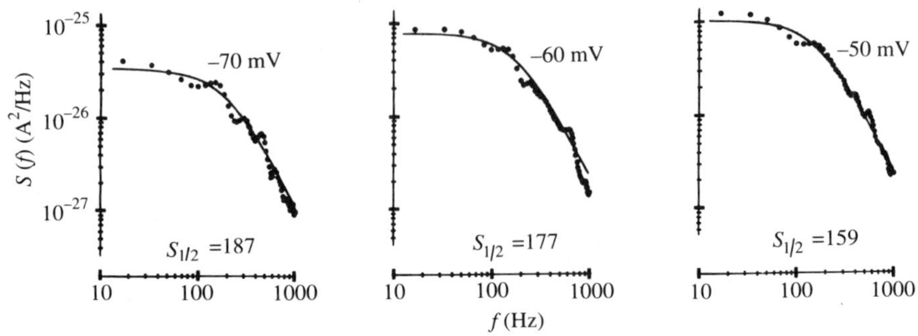

Fig. 3.14. Fluctuation spectra of sodium current flowing through calcium channels modified by a calcium-free external solution (Krishtal *et al.* 1981). The fits to Lorentz functions are indicated by continuous lines. Holding potential: −90 mV. Test potentials and cut-off frequencies are indicated near each spectrum.

achieved between the membrane surface and the internal wall of the micropipette after sucking; in this case the leakage resistance in this region increases up to 10^9–10^{11} Ohm (hence its laboratory jargon term 'giga-seal'). Such a large resistance reduces the background noise of recording by about an order of magnitude and thus enables one to record current jumps appearing during opening or closure of single ion channels (Hamill *et al.* 1981).

In the simplest form, the membrane patch sucked into the micropipette maintains its connection with the rest of the cell surface. However, its adhesion to the pipette surface is so strong that it can be torn off from the cell when pulling the pipette. After that the activity of single channels in the isolated membrane patch can be recorded. In this case the external surface of the patch turns out to face the inside of the pipette; if it has to be influenced, the influence should proceed from the pipette. The internal surface of the membrane now faces the outside, and it can also be affected ('inside-out patch').

The method has been used successfully for the analysis of behaviour of single chemically operated channels which, as already mentioned, have a large individual conductance. In fact, the conductance of unitary acetylcholine-activated channels proved to be 44 pS; the mean open time is also high (27 ms at −80 mV) (Fenwick *et al.* 1982*a*).

This method has also been used for the analysis of single calcium channels in different excitable membranes. The activity of single high-threshold calcium channels in *Helix* neurons was recorded by Lux and Nagy (1981) in conditions of increased [Ca^{2+}] at the external side of the cell membrane (50 mM). In this case rectangular current pulses of 0.3–0.6 nA could be recorded in response to depolarization. The frequency of steps sharply increased with the increase of depolarization. The mean open time of these channels was about 3 ms; however, this value was little dependent on the membrane potential (as follows from the analysis of fluctuations). The same peculiarities of the kinetics were characteristic of the channels carrying Ba^{2+} ions, but in this case the current through the single channel showed about a twofold increase. In the course of calcium current inactivation the main change was a prolongation of the mean closed time; the mean open time was little affected. The current value through single channels was not changed. Similar data were obtained by Brown *et al.* (1984) and Kunze *et al.* (1984).

A detailed analysis of these currents has shown that two different closed times, short and long ('fast' and 'slow'), are characteristic of the corresponding channels. Mean short closed time was about 1.8 ms; it was little dependent on temperature. Mean long closed time increased from 4.9 ms to 11.0 ms upon lowering of the temperature by 20 °C. Mean open time also proved to be highly dependent on temperature (Lux and Brown 1984).

The activity of single high-threshold calcium channels has been analysed on membrane patches from *Limnea* neurons by Kostyuk *et al.* (1985*b*). According to our data, the use of cells from these freshwater molluscs had a considerable technical advantage over the cells of *Helix* due to much easier treatment of the membrane surface (in freshwater molluscs the connective sheath around the neurons is much softer than that in terrestrial ones; therefore, the giga-seal was more perfect in this case). The mean value of current pulses was 0.7 ± 0.1 pA if the external solution containing 40 mM Ba^{2+} was used and the testing membrane potential was 5 mV.

Thus, high-threshold currents measured directly in single calcium channels are actually several times larger than those calculated by current fluctuation spectra; it is not clear why these calculations gave underestimated values. Examples of barium currents through single calcium channels in the somatic membrane of a *Limnea* neuron are presented in Fig. 3.15.

It should be noted that it is very difficult to obtain reliable giga-seals, even in the neurons of freshwater molluscs, and successful recording of the calcium channel activity was possible only in rare cases. A more convenient object in this respect proved to be mammalian nerve cells (mouse), especially of new-born animals (Shuba and Teslenko 1987; Kostyuk *et al.* 1988*a*). The external surface of the membrane seems to be much 'cleaner' during isolation of such cells and it, therefore, seals easily with the recording pipette.

When recording currents from the membrane patch in such neurons it was noted that, despite the presence of a well-expressed integral current, it was possible to record single-channel currents only in a small number of membrane patches (5–10 per cent). This finding indicates an extremely uneven distribution of such channels on the cell surface. The problem is also complicated by the fact that the function of the channel, if it is available in the

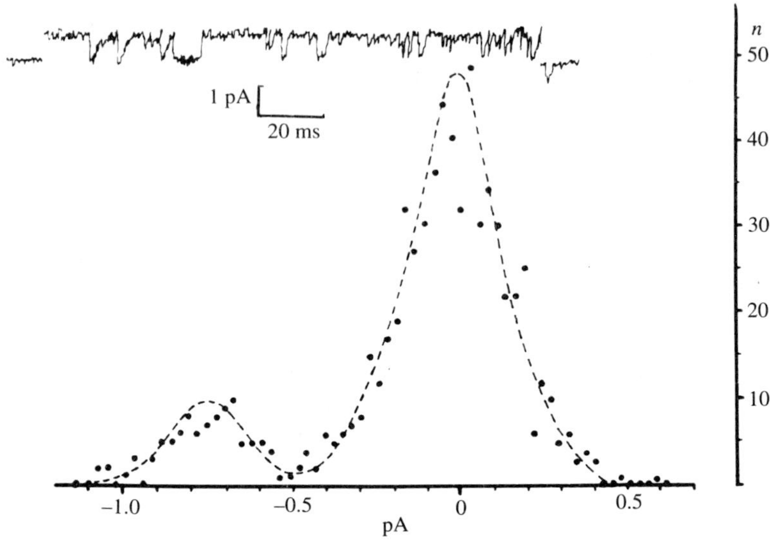

Fig. 3.15. Amplitude histogram of barium current through a single calcium channel in a snail neuron (Kostyuk *et al.* 1985*b*). On the inset one of the actual current records is presented.

membrane patch, is uneven, i.e. the active periods can be changed by long 'silent' periods. Therefore, to raise the effectiveness it is necessary to record data continuously in analog form on a magnetic tape and then to perform their digitization after finishing the experiment, for further processing by a computer. The algorithm of the processing consists of the following main blocks. The initial data are sorted according to the presence or absence of events. Realizations 'without events' are averaged and then the averaged realization is subtracted from each realization with an event. This procedure allows one to remove almost completely the distortions caused by capicitance artefacts, incomplete leakage compensation and others. Then a zero current line and a reference line of events located at 3–5σ below the zero line are determined for each realization. This line cuts off the bulk (99.73 per cent) of the background noise. All current pulses crossing this line are considered to be linked with the activity of single channels. At a half level between zero and maximal current lines the so-called line of transition into a closed state is being set up. Although the location of this line is chosen somewhat arbitrarily, its introduction in the algorithm of processing greatly expands the possibility of measurements of shorter values of closed states of the channel (as, due to restriction of the pass band of the recording system, the current has not always time to reach zero value at fast closure of the channel). Using these lines, each realization acquires an ideal form that helps one to calculate distribution histograms of open and closed states, duration of bursts in the channel work, and delay until its first opening.

In order to record selective activity of calcium channels, the solution in the pipette has to be chosen so that the currents through other types of ion channels are removed. Since during potential recording from the membrane patch it is difficult to control the membrane potential, its reduction to zero can be performed by using a high-K^+ extracellular solution and then a desirable value is set up by a voltage-clamp system.

The current pulses recorded from rat sensory neurons showed a different unitary amplitude, which agrees with the conclusion, made from the analysis of microcurrents, about the separation of calcium currents into several types. The detailed analysis of recordings revealed three types of channels. Figure 3.16 demonstrates all three types of the activity of single calcium channels when strontium ions were used as current carriers. The first type (Fig. 3.16a) could be observed in about 25 per cent of the neurons studied. Current pulses related to the functioning of this type of calcium channels were activated at a potential range of between −60 and −20 mV and revealed a clear potential- and time-dependent inactivation [low-threshold inactivating (LTI) channels]. The channels of the second type could be separated into two subtypes. Those of the first subtype needed depolarization to about 0 mV for activation and also revealed time- and

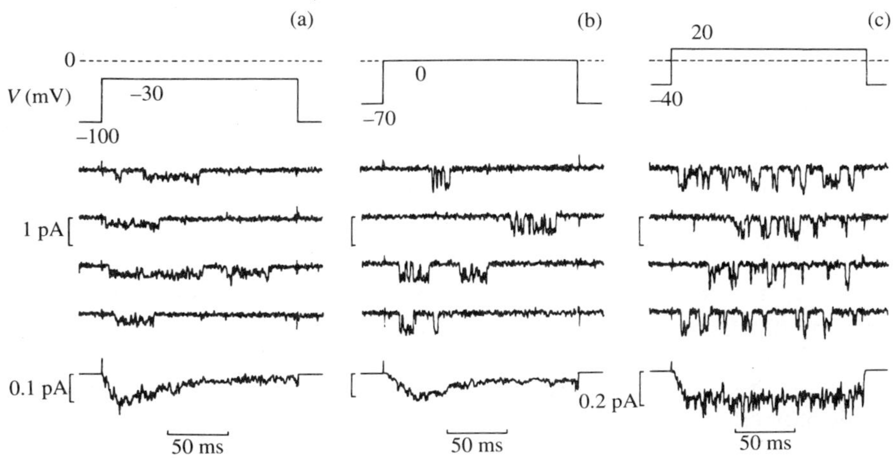

Fig. 3.16. Three types of single calcium channel activity in the membrane of a mouse sensory neuron (Kostyuk *et al.* 1988*a*). Records obtained in cell-attached configuration at holding and test potentials are indicated in the pulse protocol shown above. Lower traces show the corresponding averages of the current records. $[Sr^{2+}]_{out}$: 60 mM. LTI (a), HTI (b), and HTN (c) channels are presented.

voltage-dependent inactivation [high-threshold inactivating (HTI) channels; Fig. 3.16b]. They quickly disappeared in membrane patches torn off the cell and were slightly sensitive to organic blockers (for instance, D-600 suppressed them by about 10 per cent). Channels of the second subtype were completely activated only at depolarizations up to +20 mV and showed only negligible inactivation at long depolarizing shifts [high-threshold non-inactivating (HTN) channels; Fig. 3.16c]. Their function was especially rapidly disturbed in torn-off membrane fragments and was the most sensitive to the action of organic antagonists and agonists.

Table 3.1 compares the conductances of these channel types when different ions were used as current carriers. As is seen, they differ significantly; the most highly conducting are HTN channels and the least-conducting are LTI channels (cf. also Fig. 3.17). Furthermore, a definite difference is observed in the selectivity of different channel types for divalent cations; if HTN channels pass Ba^{2+} ions much better than Ca^{2+} ions, LTI channels do not significantly show such selectivity. More detailed separation of the peculiarities of the functioning of the three types of calcium channels could be performed from statistical analysis of the distributions of their open and closed states.

Table 3.1. Conductances of three different types of Ca channel in dorsal root ganglion sensory neurons

		Channel type		
		LTI (T)	HTI (N)	HTN (L)
Activation threshold (mV)		−70	−40	−30
	Ca	6.3	7.2	7.6
Conductance (pS)	Sr	5.7	7.2	9.0
	Ba	7.2	11.4	18.4

The open-time histogram for HTI channels could be well approximated by one exponent, the time constant (τ_{op}) which was about 1.2 ms, and was little dependent on membrane potential. The closed-time histograms had a more complex character and included the main initial exponential component (little dependent on the potential; its time constant $\tau_{cl}^{(f)}$ was about 0.9 ms) and a low-amplitude 'tail' up to 100 ms in duration (Fig. 3.18). The presence of such a tail indicates the distribution of channel-open times in groups ('bursts') separated by periods of prolonged 'silence' of the channel. For reliable separation of such bursts a simple procedure has been proposed, according to which the intervals $0 \leqslant t \leqslant 3\tau_{cl}^{(f)}$ were ignored (such intervals were considered as channel closures inside the burst). Thus obtained, the burst

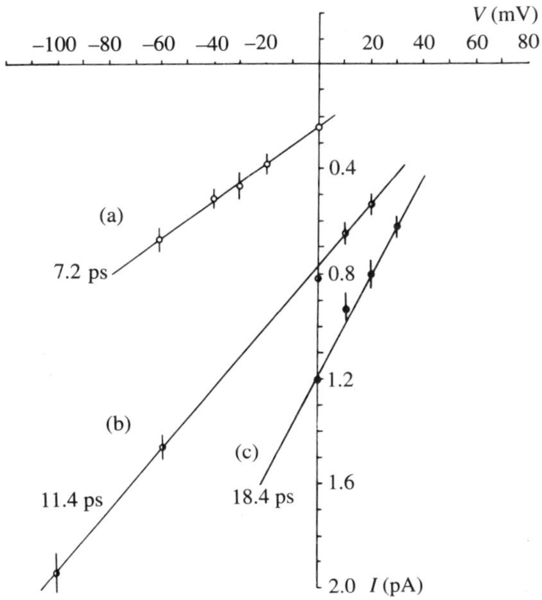

Fig. 3.17. *I–V* relations for LTI (a), HTI (b), and HTN (c) channels (Kostyuk *et al.* 1988*a*). The resting potential was zeroed using a high external $[K^+]$. Data were collected from several patches. The mean unitary conductance is indicated near each curve.

distribution histogram could be approximated by two exponentials—a fast potential-independent one with the time constant $\tau^{(f)}_{burst}$ of about 0.6 ms, and a slow, strongly potential-dependent $\tau^{(s)}_{burst}$, the time constant of which increased from 4 ms at -15 mV to 20 ms at $+25$ mV. The fast component obviously represents short openings of the channel not united in bursts, and only the slow component represents true bursts.

Apart from sensory neurons, it was very convenient to study LTI channels on non-differentiated neuroblastoma cells, in which these channels are the sole mechanism for the creation of a calcium conductance. Distribution of open states in these channels also obeyed a mono-exponential law; the time constant of such a distribution was the largest, somewhat decreasing with an increase of depolarization (from 2.0 to 0.8 ms). Open states also tended to be grouped into bursts whose duration sharply increased with an increase of depolarization (Kostyuk *et al.* 1988*a*, *b*).

A peculiar phenomenon has been described for single HTI channels. Apart from bursts of opening, during which the current amplitude reached a maximal value and then vanished to zero at closures of the channel, there

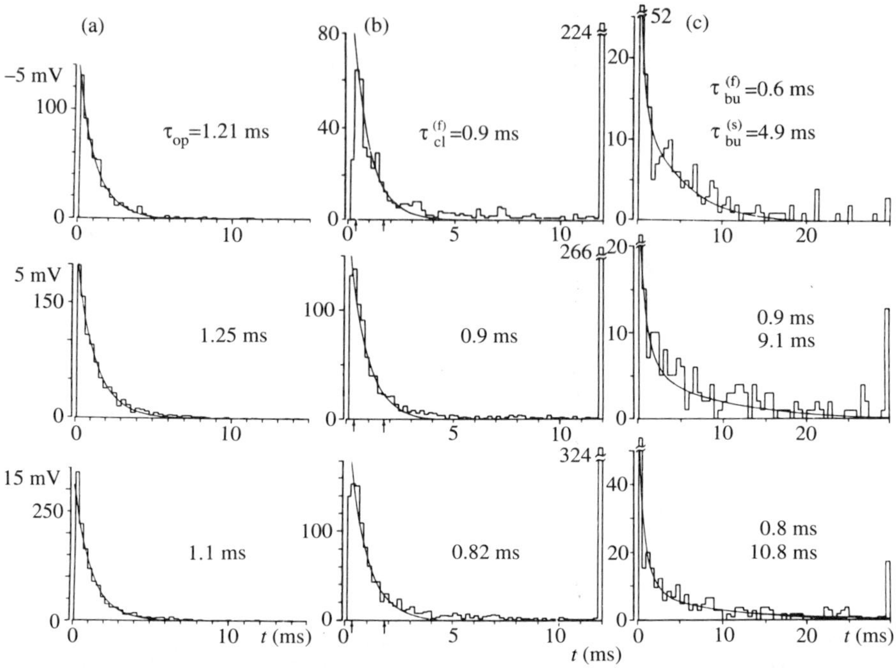

Fig. 3.18. Kinetic characteristics of single HTI channels (Kostyuk *et al.*, 1988*b*). Open-time (a), closed-time (b), and burst-length (c) histograms are presented. The corresponding time constants are indicated near the histograms; in (c) the distribution is approximated by two exponents. Records were taken at three different test potentials (as indicated on the left).

were bursts (about 10 per cent of the total number) during which the conductance attained an intermediate level (Shuba and Savchenko 1987). Intermediate values for current amplitude could be observed in all cells under investigation, and within the whole range of membrane potentials at which adequate current recording was possible. Figure 3.19 demonstrates selected current traces for three values of membrane potential in which this tendency revealed itself especially clearly. It can be seen that various transitions could occur between different levels of conductance: the channel could pass from the closed state to the state corresponding to both maximal conductance and to a sublevel of conductance. When the current reached maximal amplitude in the pulse, the channel could return to the closed state also during partial conductance. Transitions from the state of partial conductance to either the completely non-conducting state or to the state of maximal conductance were also possible.

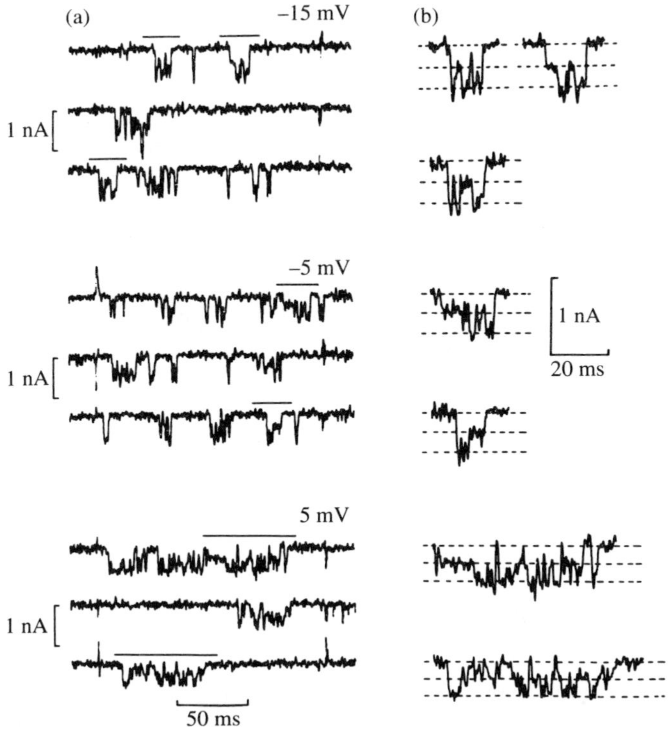

Fig. 3.19. Two conducting states of a single HTN channel (Shuba and Savchenko 1987). Bursts of channel openings shown in (b) are taken from part of the records in (a) marked by continuous lines. Test potentials are indicated near each group of records.

Channel conductances in the main and intermediate states, found by the slope of current–voltage curves for these states (see Fig. 3.20), were 7 and 5.7 pS, respectively (with 60 mM calcium in the medium). Unfortunately, it was not possible to evaluate reliably the calcium channel lifetime in the intermediate open state because of insufficient resolving power of the recording system. Approximately, one can say that this parameter changes within the same range as the lifetime in the same open state.

The results cannot be explained by the existence at the site of recording of two different channels with different current amplitudes, or of two similar channels which occasionally operate synchronously; it is also impossible to explain these data by frequency limitations of the recording system. Therefore, it is most likely that HTI channels can exist in two open states—in the main state and in an additional state.

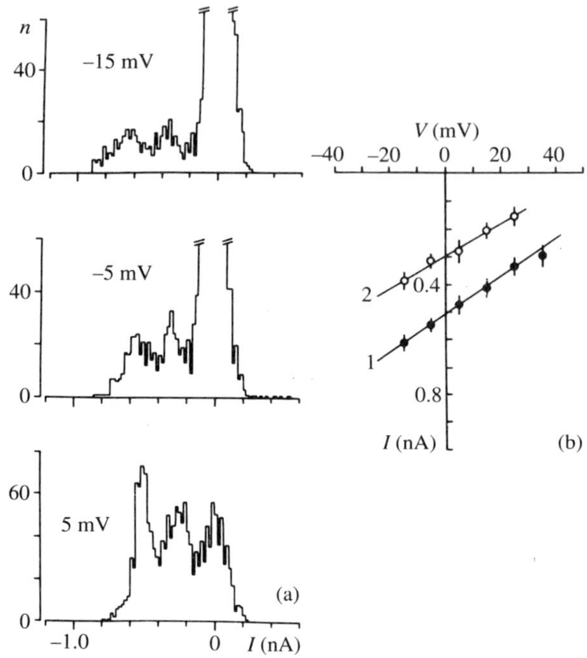

Fig. 3.20. Amplitude characteristics of the two conducting states of a single HTN channel (Shuba and Savchenko 1987). Amplitude histograms (a) of currents recorded at three test potentials, and *I–V* curves for the main (1) and intermediate (2) conducting states are shown.

The findings about the existence of conductance sublevels have also been obtained for other types of ion channels. Thus, such sublevels were especially revealed in the channel operated by the nicotinic cholinoceptor of embryonic cells, which may be related to metabolic instability of cholinoceptors in developing tissue (Colquhoun and Sakmann 1981). In mature tissue, the relative number of bursts of pulses containing conductance sublevels was reduced to 1–2 per cent. Since experiments in our laboratory have been performed on embryonic sensory neurons; it can be suggested that in adult cells this phenomenon will also be insignificant.

The kinetic characteristics of HTN channels could be most successfully analysed on phaeochromocytoma PC-12 cells, in which the calcium current is created almost exclusively by this type of channel (Shuba and Lyubanova 1988). The results obtained on phaeochromocytoma cells showed that these HTN channels had exactly the characteristics of such channels in sensory neurons.

The most characteristic feature of this channel type was the fact that its open-time histograms could be well fitted only by a bi-exponential function (Fig. 3.21). A fast component was slightly dependent on the potential, decreasing from 0.5 to 0.2 ms with an increase of depolarization of 30 mV; a slow component was practically unchanged and its time constant was about 1.8 ms. Obviously, this property is connected with the ability of the channel to acquire two different conducting states with different lifetimes. Other kinetic properties of this channel type did not substantially differ from those of HTI channels; their openings were also grouped into bursts whose duration strongly depended on membrane potential level.

Similar results on the kinetics of different types of single channels in sensory neurons and in clonal cells have been obtained independently in other laboratories. In Tsien's laboratory the following designations have been proposed for their identification: T (LTI), N (HTI), and L (HTN) channels (see Tsien *et al.* 1988). Three channel types have been described in

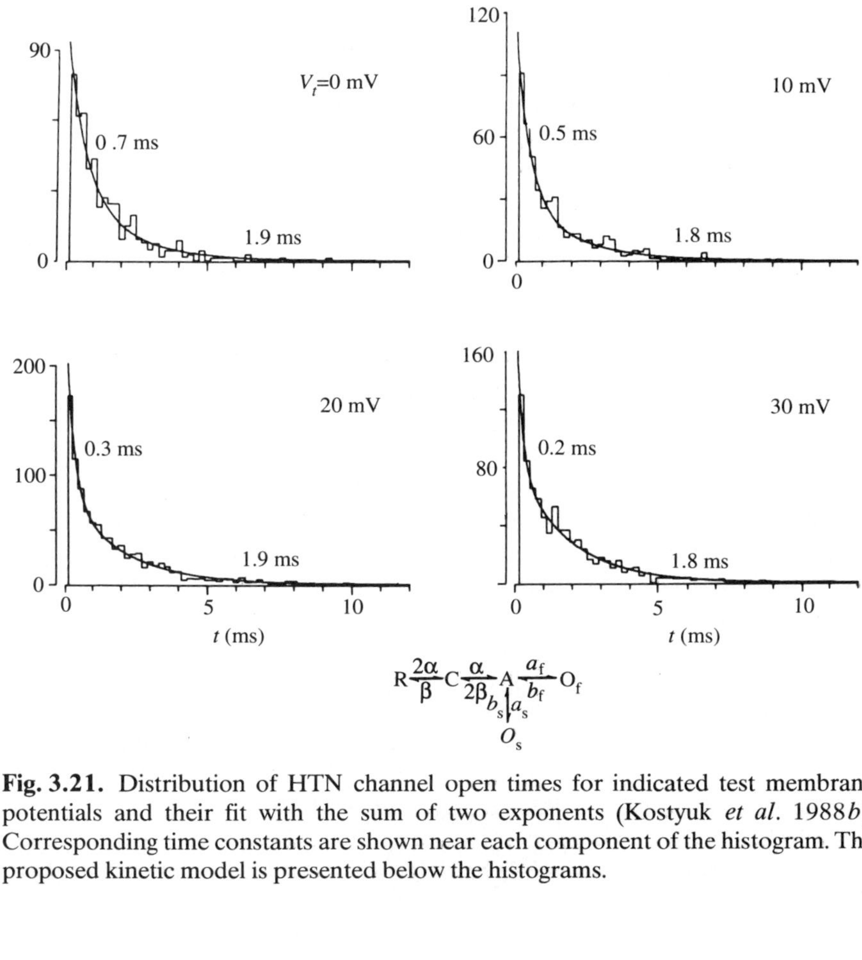

Fig. 3.21. Distribution of HTN channel open times for indicated test membrane potentials and their fit with the sum of two exponents (Kostyuk *et al.* 1988*b*). Corresponding time constants are shown near each component of the histogram. The proposed kinetic model is presented below the histograms.

mouse melanotrophs (Taraskevich and Douglas 1989). In cultured rat hippocampal neurons three types of calcium channels have been identified as low-threshold (I_l), and high-threshold fast inactivating (I_{hf}) and slow inactivating (I_{hs}) (Ozawa *et al.* 1989).

However, other studies failed to clearly separate additional groups of high-threshold calcium channels. So, sensory neurons of chick and rat (Carbone and Lux 1989*a*, *b*; Swandulla and Armstrong 1988) revealed only low-voltage activated (LVA) and high-voltage activated (HVA) calcium channels. The same is also characteristic of rat cerebellar Purkinje cells (Bossu *et al.* 1989) and differentiated N1E-115 neuroblastoma cells (Yoshii *et al.* 1988).

Kinetic models of calcium channels

The results described above indicate that application of the original Hodgkin–Huxley model to the functioning of calcium channels cannot explain a number of peculiar features of their kinetics. It fails to account for the results of spectral analysis of calcium current fluctuations, the analysis of relaxation kinetics of calcium conductance in response to small potential displacements, kinetics of current 'tails' at strong repolarization of the membrane, and especially the results of the analysis of single-ion channel activity. All these findings suggest the presence of a fast potential-independent kinetic stage preceding the open state of the channel.

The simplest kinetic model with which one can reconcile both groups of the results would be a scheme according to which the channel can pass consecutively through several closed states before it arrives to the final conducting state. Considering the good fit of the activation kinetics of calcium currents to the Hodgkin–Huxley model at the square power of the *m*-variable, it is reasonable to use the following model:

$$\mathrm{R} \underset{\beta}{\overset{2\alpha}{\leftrightharpoons}} \mathrm{C} \underset{2\beta}{\overset{\alpha}{\leftrightharpoons}} \mathrm{A} \underset{b}{\overset{a}{\leftrightharpoons}} \mathrm{O} \qquad (3.9)$$

in which the first two transitions represent the Hodgkin–Huxley model, with potential-dependent transition rates α and β formally corresponding to displacement of gating particles, and the third one represents a fast transition of the channel from an activated but still closed state (A) into an open state (O) which is little dependent on the potential. In this case $\alpha, \beta \gg a, b$. The analysis of the applicability of this model to the description of the kinetics for both microcurrents and single-channel currents was initially performed on HTI channels of rat dorsal root ganglion neurons (Shuba and Teslenko 1987); then it was shown that it is also applicable to other types of calcium channels (Kostyuk *et al.* 1988*b*; Kostyuk *et al.* 1989*b*).

As indicated above, lifetime histograms for HTI channels (in the open state) have a mono-exponential character with a voltage-independent mean time (τ_{op}). A similar feature is found for the initial component of closed-time histograms whose mean time ($\tau_{cl}^{(f)}$) is also voltage-independent and equal 0.9 ms. There are reasons to suggest that these histograms are characteristic of potential-independent conformational transitions of the channel between closed activated (A) and open (O) states (stage $A \underset{b}{\overset{a}{\leftrightarrows}} O$ which lead to a 'flickering' effect of the channel. Correspondingly, $\tau_{op} = 1/b$, and $\tau_{cl}^{(f)} = 1/a$.

The measurement of rate constants α and β from the results on the activity of single channels is more complex and is based on the results of distribution of channel-opening bursts as well as on the latent periods of the first channel opening. As was shown above, distributions of burst duration can be approximated by two exponents, of which the fast $\tau_{burst}^{(f)}$ is practically voltage-independent and the slow one, $\tau_{burst}^{(s)}$, is strongly potential-dependent. It seems that the fast component represents single-channel openings not combined in bursts and that the physical meaning of this process is the maintenance of the balance between activated (A) and open (O) states, while the slow component characterizes the lifetime of the total state $A \underset{b}{\overset{a}{\leftrightarrows}} O$. Mathematical analysis shows that from the mean time of the burst length, $\tau_{burst}^{(s)}$, and the rate constant β, the mean lifetime of the gating particle in the 'activated' state must be:

$$1/\beta = 2\tau_{burst}^{(s)}\ \tau_{burst}^{(f)}[\tau_{op} + \tau_{cl}^{(f)}]. \qquad (3.10)$$

In order to determine the rate constant α, one should have additional information about waiting-time histograms for the first channel opening after the onset of a depolarizing shift. Figure 3.22 demonstrates such histograms for HTI channels of mouse sensory neurons, obtained with increasing depolarizations. As can be seen, all the waiting-time histograms had a maximum, which shifted to shorter time intervals with increased depolarization. These histograms are very convenient for determining the rate constants of transition, because the amplitude and position of the maximum are very sensitive to change in α and β values. They can be approximated by the difference of two exponents, which are also presented graphically in the same Figure (Fig. 3.22). From their time constants τ_s and τ_{vs}, the rate constant α and mean lifetime of the gating particle in an activated state can be determined:

$$1/\alpha = 2(\tau_s \tau_{vs})^{1/2} \qquad (3.11)$$

Thus, the complete set of rate constants determining the behaviour of a single calcium channel can be obtained in the frame of the proposed generalized model. To test the adequacy of such a description of the kinetic characteristics of this channel, a comparison was made between the experimentally obtained closed-time distribution and theoretical distributions calculated from mathematical analysis of the model, according to which such a distribution must represent a sum of three exponents with time constants, $\tau_{cl}^{(f)}$,

$\tau_{cl}^{(s)}$, and $\tau_{cl}^{(vs)}$. The latter two values must be identical to the exponents describing the waiting-time distributions for the first opening. The amplitudes of the corresponding exponential components related to the amplitude of the fast component were determined according to the following equations (Shuba and Teslenko 1987):

$$A_s = \frac{2[\beta\tau_{cl}^{(f)}]^2\tau_{vs}}{(\tau_{vs} - \tau_s)(2\alpha\tau_{vs} - 1)}$$

$$A_{vs} = \frac{2[\beta\tau_{cl}^{(f)}]^2\tau_s}{(\tau_{vs} - \tau_s)(1 - 2\alpha\tau_s)} \quad (3.12)$$

The results of such a comparison are presented in Fig. 3.23, which shows experimentally obtained closed-time and interburst gap histograms and their theoretical approximations. Satisfactory agreement between the theoretical and experimental curves is seen, thus supporting the possibility of using the proposed model for the description of the kinetic behaviour of single high-threshold calcium channels. Similar relations between different components of closed-time histograms have also been reported for cardiomyocytes by Cavalie *et al.* (1986). It should be noted that as compared with the 'fast' potential-independent component, the total contribution of potential-dependent components to closed-time histograms is insignificant and hardly changes with alterations in V_m. This conclusion is in good accordance with experimental results. However, Hagiwara and Ohmori (1983) reported considerable changes in the relations between the amplitudes of potential-dependent and potential-independent components of the closed-time histograms (from 0.07 to 0.4 for a 25 mV depolarization); this discrepancy may be due to the presence in their experiments of several calcium channels in a membrane patch.

Figure 3.23 also shows that even though the theoretical distribution describes correctly most of the events in the closed-time histograms (about 95 per cent), a small proportion of them still form an extra-slow tail that cannot be predicted by the theory. Most probably, this tail is due to inactivation of a calcium channel not considered by this theory. An extension of the model to include inactivated channel states seems to be premature at present because of insufficient experimental data about inactivation kinetics of single calcium channels and the absence of clarity in the understanding of the mechanisms of calcium current inactivation (see Chapter 4). It should be borne in mind that since the identification of the potential-dependent rate constants α and β was based on the histogram of burst duration and waiting times to the first channel opening, which are least affected by the inactivation process, the distortions of the results describing activation of the calcium channel by superposition of inactivation should be minimal.

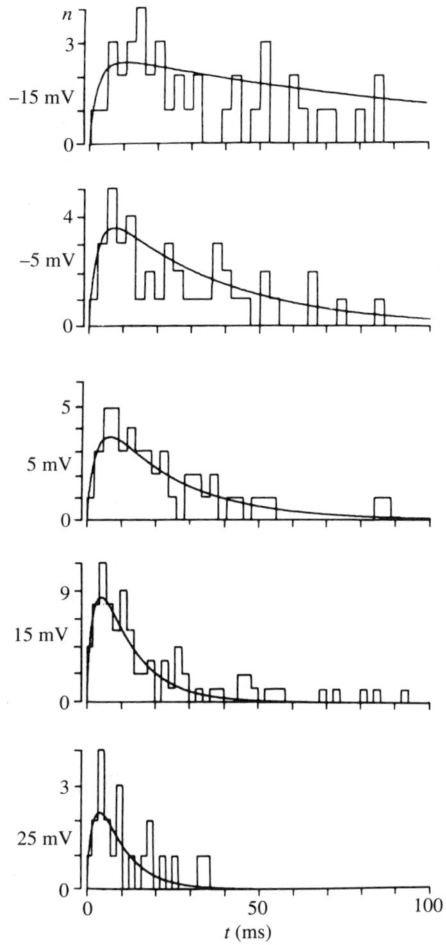

Fig. 3.22. HTI channel waiting-time histograms at different membrane potentials (indicated on the left) and their approximation by the difference of two exponents (Kostyuk *et al.* 1988*b*).

Figure 3.24 presents the determination of the mean rate constants, τ and β, responsible for the initial two stages of calcium-channel activation. The values obtained in a semilogarithmic scale fit well to the straight lines that correspond to the expected exponential dependence of these constants on V_m. The logarithm of the relation $\alpha(V)/\beta(V)$ has a slope of 2.2, representing the effective valency of a postulated single-gating m-particle in its trans-membrane transition during calcium-channel activation. As was indicated above, similar values (2.4–2.9) have been obtained in the analysis of

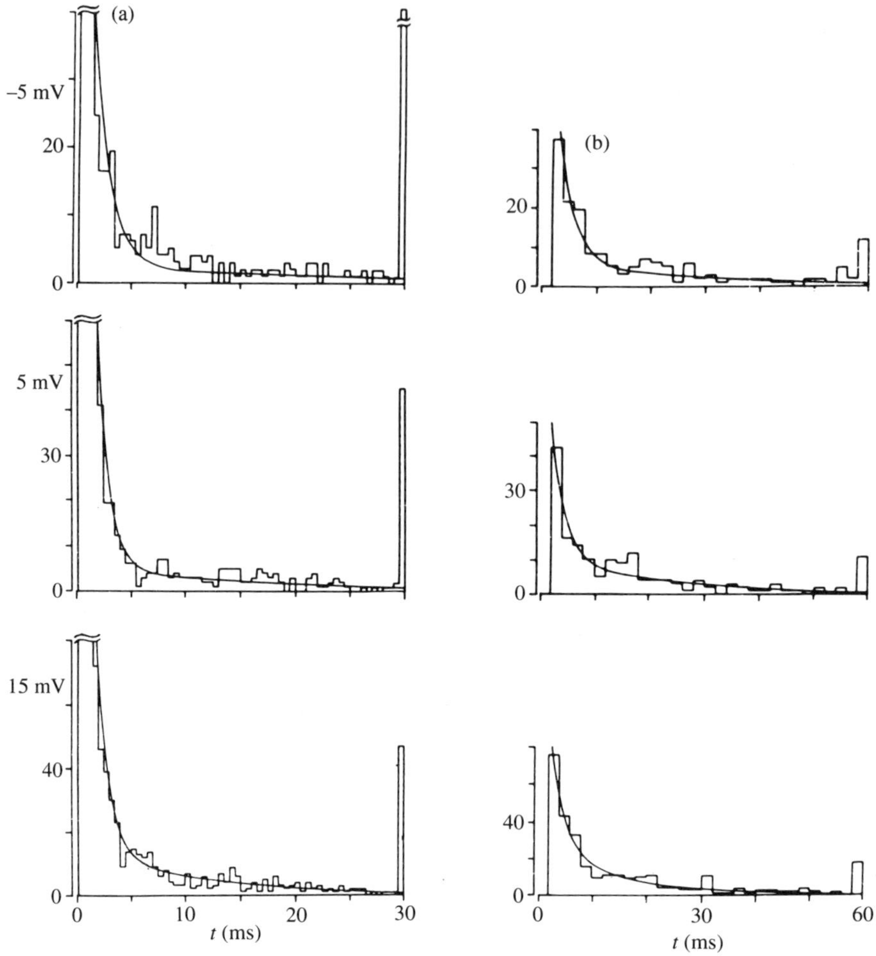

Fig. 3.23. HTI channel closed time (a) and interburst shut time (b) histograms at three different membrane potentials (Kostyuk *et al.* 1989*b*).

stationary properties of macroscopic calcium currents, as well as of asymmetric displacement currents accompanying their generation.

The rate constants obtained for the gating mechanism of the high-threshold calcium channel on the basis of the proposed model show good coincidence with the analysis of deactivation kinetics of the corresponding macroscopic currents. At very negative repolarization potentials, the potential-dependent transitions of the channel to a non-conducting state become very fast, and reverse transitions to the activated state are practically impossible. Therefore, at these potentials, the process of channel deactiva-

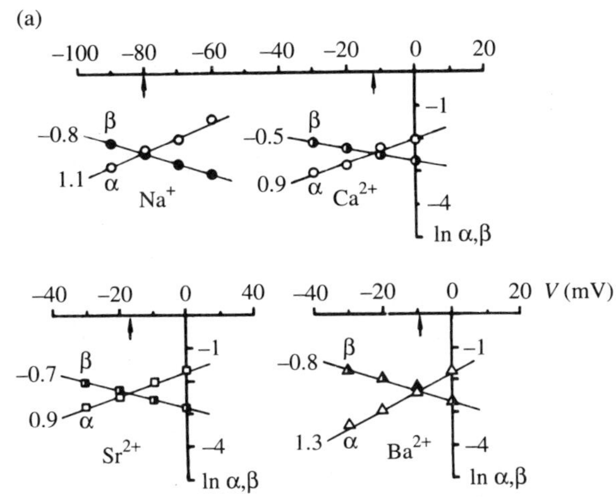

Fig. 3.24. Potential-dependence of the rate constants α and β, calculated for HTI channels of neuroblastoma cells (a) and HTI channels of mouse dorsal root ganglion neurons (b) from single channel recordings (Kostyuk *et al*, 1989*b*). Carriers and semilogarithmic slopes of α and β are shown near each graph. Arrows indicate potentials of intersection for α and β lines that correspond to half-activation of the channel.

tion should in fact be limited by the velocity of potential-independent transition of the channel A ← O. This velocity will be close to an inverse value of the mean open time of the channel. At less negative repolarization potentials the voltage-dependent transition rates between different activated states becomes important for evaluation of the deactivation kinetics of the calcium current.

At present, a tentative assessment of applicability of this model to the description of kinetic characteristics of other types of calcium channels has been performed. For LTI channels the main property of the final stage of conformational transitions seemed to be a well-expressed potential-dependence. With a stronger depolarization, the mean open time (which turned out to be more prolonged as compared with that for other channel types) shortened noticeably, as can be seen from plots presented in Fig. 3.18. Calculations of kinetic constants α and β for LTI channels showed only insignificant quantitative differences from these values for HTI channels, as well as their similar potential-dependence (Fig. 3.24).

LTI channels proved to be a convenient object for more detailed analysis of the changes in the functioning of the gating mechanism which occurs upon a change of permeant ion species in the channel. Replacement of some divalent cations by others provoked some changes in all kinetic constants of the channel model under consideration. As is seen from the plots in Fig. 3.24, this entailed a change in the slope of semilogarithmic dependences of α and β constants on the potential as well as in their relation $\alpha(V)/\beta(V)$. Especially marked were the changes in the functioning of the gating mechanism when calcium channels were modified into sodium ones by removal of divalent cations from the medium. Apart from a considerable shift in the voltage-dependence of the kinetic characteristics, qualitative changes also took place. The most important feature was the dependence of the kinetic characteristic of these channels on the concentration of sodium ions when they pass through the channels, i.e. on their flux through the channel (see Fig. 3.25). In order to explain such a dependence, a suggestion has been put forward about electrostatic interaction of permeating ions with the channel wall (Kostyuk *et al*. 1988*b*). It is supposed that transition of an ion through an open channel produces local displacements of charged groups forming the steric region, which are dependent on the ionic radius ρ_i. This leads to the formation of dipoles along the ion trajectory which exist during the time of dipole relaxation $\tau \approx 10^{-7}$ s. The mean induced dipole moment, d_i, depends on the ion transfer rate through the channel, ω_i. In the case of a stationary ionic current these values can be determined as follows:

$$\omega_i = (kT/e)(g_i/z_i^2 e) \tag{3.13}$$
$$d_i = \omega_i \tau e \delta_i / (\omega_i \tau + 1)$$

where $kT/e = 0.025$ V.

One may assess possible d_i values, taking into account that when passing through the steric region a permeant ion is surrounded by three water molecules having an effective diameter of 3 Å. For divalent cations the values of induced dipole moments must be small and they cannot strongly affect the voltage-dependence of channel closure. Obviously, the same is true for lithium ions which pass through the channel without steric difficulties. A different situation is observed if Na^+ is used as a permeant cation. If the conductance of the modified channel for Na^+ ions is 10–100 pS (depending on the concentration of ions and presumed displacement of charged molecular groups in the selectivity filter of 0.2 Å), the ion transfer rate will reach 10^6–10^7 s^{-1}, i.e. correspond to dipole relaxation frequencies. Therefore, the contribution of dipole moments (0.1–0.5 D) induced by the sodium flux must be considerable. At large sodium concentration the changes should be so important that mean channel open time becomes almost potential-independent. After channel transition from the open state to the nearest closed state, the steric region of the channel becomes narrow and ion flux through it is no longer possible. That is why the mean intraburst closed time, $\tau_{cl}^{(f)}$, which characterizes the channel 'reverse' transition to an open state, is practically independent of the permeant ion species and concentrations.

Nevertheless, in the channel closed state the ion species can also affect kinetic characteristics of the channel. As was shown above, variation in the

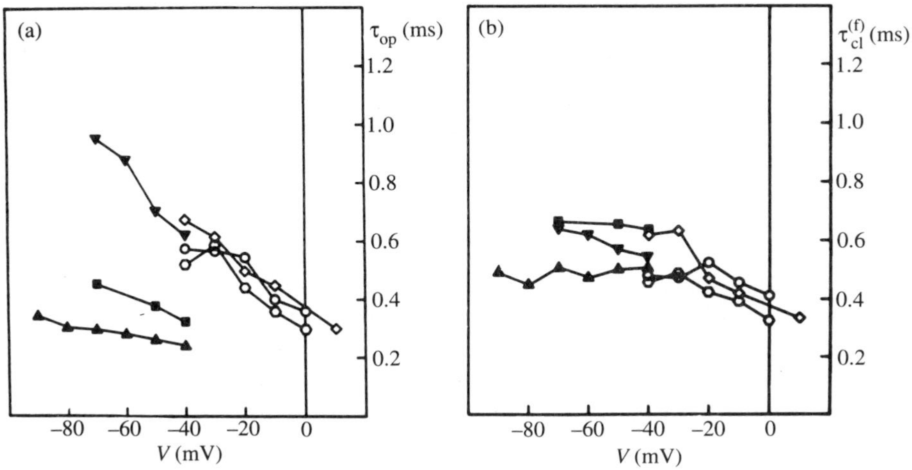

Fig. 3.25. Potential-dependence of rate constants of single LTI channels (Kostyuk *et al.* 1988*b*). (a) Mean open time; (b) mean time of intraburst gaps. Charge carriers were: Ca^{2+} 60 mM (○), Sr^{2+} 60 mM (⬡), Ba^{2+} 60 mM (◇), Na^{+} 100 mM (▲), Na^{+} 40 mM (■), Li^{+} 100 mM (▼).

permeant ion type results in changes in the logarithmic slopes of $\alpha(V)$ and $\beta(V)$ and in the shift of the half-activation potential of the channel. These phenomena can be explained by supposing that channel binding sites have a geometrical arrangement along the pathway of gating charge transfer. Electrostatic interaction between the bound cation and the gating charge will differ, depending on cation radius, its chemical affinity to the binding site, and its valency. A high-affinity binding site in the calcium channel outer mouth, responsible for modification of selectivity, can add much to such an interaction if the initial localization of charged gating particles in the channel resting state is in close proximity to this binding site. The 70 mV hyperpolarizing shift of the activation curve for the modified calcium channel, which is much larger than the possible shift due to an increase in surface potential in the absence of divalent cations (about −40 mV, see Chapter 2), indicates that this might be the case. More detailed discussion of possible mechanisms of the influence of permeant cation type on initial stages of conformational transitions in calcium channels is premature, due to the lack of sufficient data about their molecular structure.

The modelling of the influence of permeant ion type on the channel gating mechanism has been performed only for LTI channels. Based on general considerations, one may think that the above suggestions could be true for other types of calcium channels.

Finally, the most essential property of the gating mechanism for HTI channels (N-type) is its ability to transform the channel into two different conducting states, significantly differing from each other by their mean open times. In the framework of the model used, it can be suggested that these states are involved in the last kinetic activation stage as follows:

$$O_s \underset{b_s}{\overset{a_s}{\leftrightharpoons}} A \underset{b_f}{\overset{a_f}{\leftrightharpoons}} O_f \tag{3.14}$$

where O_f are short-living and O_s are long-living open states. However, this suggestion is preliminary and requires a more thorough analysis. The factor that determines the possibility of channel transition to either of these states is not known so far. Some organic agonists are reported to affect selectively one of these states, considerably increasing its mean lifetime. According to Hess *et al.* (1984), Bay K 8644 is an agonist of only the long component and has no effect on the short component of the calcium-channel open state in ventricular heart cells. Our data obtained on HTN channels of phaeochromocytoma cells with the use of this substance demonstrated a change in the 'fast' state, whose characteristic time increased by about 2.5 times, the 'slow' component being practically unchanged (Kostyuk *et al.* 1988*b*).

Conclusion

Like other voltage-operated channels, calcium channels in various excitable membranes have a peculiar functional component that allows them to perceive changes in the transmembrane electric field and to pass from the closed (non-conducting) to the open (conducting) state. This component is not connected rigidly with the part of the channel that determines its selectivity, since the kinetic functioning of the channel is little affected by considerable modifications of the latter. Activation kinetics of calcium currents can be described statistically by the Hodgkin–Huxley model; in most cases the most adequate assumption is a square power of the *m*-variable. The kinetics of the intramembrane asymmetrical charge displacement ('gating currents'), which accompanies activation and deactivation of calcium currents, also fits this model. However, analysis of calcium current fluctuations and of the functioning of single calcium channels shows that conformational changes occurring during activation are more complex and that the channel transition to an open state is a fast process with little potential-dependence. Different types of calcium channels (low-threshold, high-threshold inactivating and non-inactivating) seem to possess a similar gating mechanism, which differs only in the potential-dependence of the transition rate constants between different states and their quantitative values.

4. Inactivation of calcium channels

History

Inactivation of voltage-operated ion channels is phenomenologically a well-known process, which manifests itself in a gradual decline of the current during sustained membrane depolarization. This phenomenon has been studied in detail on sodium channels of the squid axon and frog nerve fibre membranes (reviewed by Meves 1978). Quantitatively, inactivation is taken into account by the introduction, in the corresponding equations, of the variable h (see Chapter 3). Another manifestation of inactivation is the stationary diminution of the maximal amplitude of the ionic current elicited by a test depolarization when the membrane potential was initially held at a decreased level. The dependence of this diminution on the holding potential level has an S-shaped form and is described as the steady-state inactivation function, $h_{\infty}(V)$.

An important point for the understanding of the nature of potential-dependent inactivation is the fact that certain factors on the internal side of the cell membrane can substantially change it without simultaneously affecting the activation of ion channels. Especially effective is the treatment of the internal side with pronase—in perfused squid axons this removes the inactivation of sodium currents completely (Armstrong *et al.* 1973). The removal of inactivation could also be achieved by a shift of pH_{in} in the alkaline region and by the action of trypsin-specific reagents (Brodwick and Eaton 1978). There are also many pharmacological substances that affect only inactivation, mostly slowing down or abolishing it (reviewed by Khodorov 1979).

On the basis of these findings, it is usually accepted that inactivation of ion channels by membrane depolarization is due to the presence in the channels of special 'inactivation gates'. As the inactivation can be most effectively modified from the interior of the cell by proteolytic agents, it may be produced by a peptide group situated at the internal mouth of the channel and capable of closing it from inside. However, this idea is certainly a simplification of reality, as more detailed investigations have shown a quite complicated relationship between activation and inactivation of sodium channels; the inactivation itself seems to be a complicated process, including a rapid and a slow component. Nevertheless, the presence in ion channels of a special mechanism which can switch-off the channel in response to membrane depolarization but not open it, is a fact of basic importance, and since the early investigations of calcium channels, this question has attracted much attention.

The analysis of inactivation of calcium channels is more complicated than that of sodium channels. First of all this is due to a slow development of calcium currents that coincides with a parallel activation of outward potassium currents. The decline of the calcium current can therefore be apparent, simulated by a superposition of a rising outward current. Several mistakes caused by this effect can be found in the literature, and reliable separation of outward currents is obligatory in the analysis of calcium-channel inactivation.

Calcium-dependent inactivation of calcium channels

In the early measurements of calcium currents, made on perfused neurons of the snail *Helix pomatia*, it was shown that the time course of their inactivation is substantially different from that of sodium currents (Kostyuk and Krishtal 1977*a*). Calcium current decline during sustained depolarization developed very slowly (over many seconds). It had a complicated time course and could be approximated by at least two exponents, with time constants of several tens and several hundreds of milliseconds (Shakhovalov 1977). The $h_\infty(V)$ function was shifted considerably in the depolarizing direction, and inactivation never became complete, even with very strong depolarizations. Because of the detection in the same experiments of the blocking effect of elevated $[Ca^{2+}]_{in}$, a suggestion was made that inactivation of calcium channels may have a special mechanism and might be induced by the recurrent blocking action of the calcium ions that have entered (Kostyuk and Krishtal 1977*b*). This suggestion has been confirmed in double-pulse experiments on *Aplysia* neurons, in which changes of a test calcium current have been measured after a conditioning depolarizing shift of different amplitude. The change of the testing current repeated the current–voltage curve of the calcium current induced by the conditioning shift. A parallel recording of calcium influx into the cell with aequorin has shown that an increase in $[Ca^{2+}]_{in}$ does indeed develop together with inactivation of the testing current, the amplitude of which depends on $[Ca^{2+}]_{out}$ and duration of the conditioning depolarization. If EGTA was injected into the cell, or after replacement of Ca^{2+} by Ba^{2+} in the extracellular solution, the depression of the testing current diminished (Tillotson 1979; Eckert and Tillotson 1981). Similar results have been obtained for the inactivation of calcium currents in *Paramecium* (Brehm *et al.* 1980). All these data definitely show that inactivation of calcium channels can occur by a mechanism that is not present in sodium channels, i.e. by the inward flux of calcium ions induced by conditioning depolarization ('current-dependent' inactivation, contrary to the above-described 'voltage-dependent' inactivation).

A confirmation of the dependence of calcium-channel inactivation on the intracellular calcium concentration has been obtained by Standen (1981) and Plant and Standen (1981) using snail neurons. The decay of calcium current in the experiments could be approximated by a single exponent, the time constant of which (in the range of several tens of milliseconds) depended on the concentration of Ca^{2+} in the external solution—it slowed down in a low-calcium solution. The recovery of the initial current amplitude occurred very slowly and with complicated kinetics (the initial phase with a time constant of 120 ms, and the late one with a time constant of about 9 s). If Ca^{2+} ions were directly injected into the cell, the calcium currents were also rapidly depressed and recovered with the time constants of 5–10 s.

However, such a simple conclusion met difficulties. It did not explain the complicated time course of inactivation in many neurons (already mentioned), which has also been observed by Brown *et al.* (1981) in *Helix aspersa* and Aschcroft and Stanfield (1982) in the insect *Carausius morosus*. The idea emerged that voltage-dependent inactivation may also exist in calcium channels besides the calcium-dependent inactivation. Thus, Brown and co-workers, having separated two components in the time course of inactivation, observed a slow down of the fast component (time constant of 20–80 ms) after injection of EGTA into the neuron; the slow component (time constant of 600–1200 ms) did not change. They also recorded some steady-state inactivation of calcium currents even at a holding potential of −75 mV, when calcium influx into the cell did not occur. After substitution of Ba^{2+} for Ca^{2+} in the external solution, the inactivation remained, although in a very slow form; intracellular introduction of Ba^{2+} did not exert any effect on it. All these findings were interpreted as an indication of the presence of both forms of inactivation in calcium channels of snail neurons.

In our experiments on perfused neurons of *Helix pomatia* we investigated the same problem in conditions of maximal elimination of possible outward currents, and reached the opposite conclusions (Doroshenko *et al.* 1984*a*).

1. The slow component of inactivation (with a time constant of several hundred milliseconds) demonstrated a potential-dependence typical of a calcium-induced process (the highest velocity at potentials corresponding to the peak of the current–voltage characteristics); the fast component did not show any relation to the latter.
2. The fast component could be almost completely abolished by an increase in pH_{in} from 7.3 to 8.2. If the current recorded at pH 8.2 was subtracted from that recorded at pH 7.3, the difference curve revealed a fast activating outward current component resembling the already described 'non-specific', or proton, outward current (see Fig. 4.1).
3. If the calcium current was blocked by calcium-channel blockers (verapamil, Co^{2+}), such an outward current, with an activation time course close to the time course of the fast phase of I_{Ca} inactivation, could

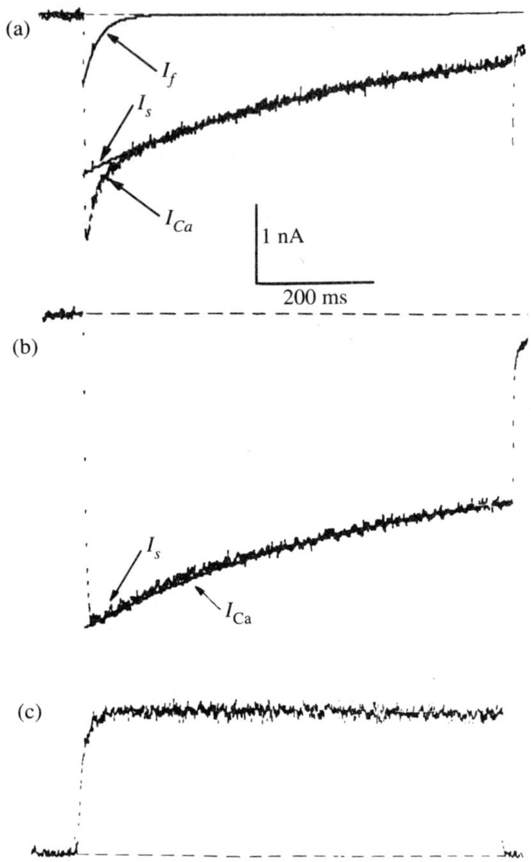

Fig. 4.1. The effect of elevated pH_{in} on the calcium inward current in a snail neuron (Doroshenko *et al.* 1984*a*). (a) Calcium current obtained at pH_i 7.3 and the approximation of its decline by two exponents; (b) current obtained at pH_i 8.1; (c) result of subtraction of (b) from (a), representing the blocked non-specific outward current. Holding potential: −50 mV.

be recorded directly, in most cases, despite the removal of K^+ ions from the cell during perfusion.

Thus, it seems that even intracellular perfusion does not guarantee a complete switch-off of all outward currents, and a residual component, carried probably by protons, may simulate fast inactivation of calcium currents. If the latter possibility is taken into account, no voltage-dependent component in the inactivation of calcium current in snail neurons can be detected.

Figures 4.2 and 4.3 illustrate some feature of such inactivation. Preliminary introduction of calcium-chelating substances into the cell should increase its calcium-buffering capacity and decrease the acting concentration of injected Ca^{2+} ions; in fact, after such injection the time constant of inactivation increased from 210 to 520 ms (Fig. 4.2). Replacement of penetrating Ca^{2+} by other divalent cations with less effective recurrent action on calcium channels (Ba^{2+}) induced a similar slow down of inactivation (Fig. 4.3). Incidentally, such a replacement also helped to abolish the fast phase of calcium current decay, presumably via an additional block of proton currents by intracellular Ba^{2+}.

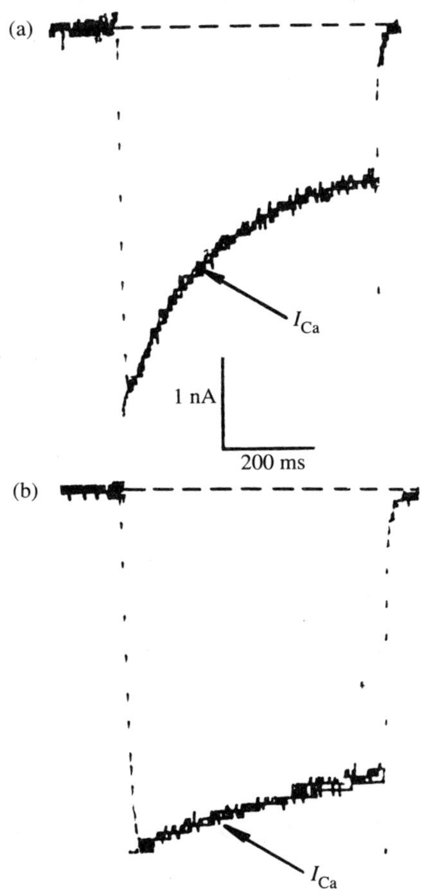

Fig. 4.2. Changes in the kinetics of the calcium current decay after introduction of EDTA (10 mM) in a snail neuron (Doroshenko *et al.* 1984*a*). (a) Before introduction; (b) after the introduction. In both cases the decay could be approximated by a single exponent.

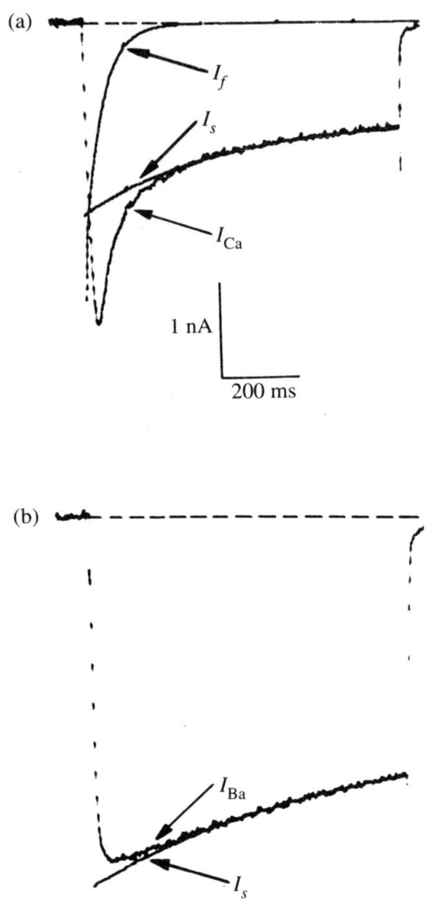

Fig. 4.3. Changes in the kinetics of the calcium current decay after equimolar substitution of Ba^{2+} for Ca^{2+} in the extracellular solution (Doroshenko *et al.* 1984*a*). (a) Ca^{2+} (decline could be approximated by two exponents); (b) Ba^{2+} (fast component practically absent).

The calcium-dependent inactivation of calcium currents in snail neurons has also been confirmed in the investigations of their 'tail currents' (Eckert and Ewald 1983*b*). The amplitude of calcium tail currents also decreased on the background of conditioning depolarization, being maximal between +10 and +40 mV, i.e. in the region of the highest influx of Ca^{2+} into the cell. Direct injection of Ca^{2+} into the neuron decreased tail-current amplitude, while EGTA injection restored it and accelerated the recovery from inactivation. The authors mentioned that the injection of chelates did not completely remove the inactivation of tail currents; this may be due to the slowness of the process of Ca^{2+} binding by EGTA, which takes seconds—much more than the inactivation decay of calcium current.

Voltage-dependent inactivation of calcium currents

The analysis of the inactivation of calcium current in neurons from other animals has been complicated by the fact that in these animals calcium is usually carried through different types of voltage-operated calcium channels. In mammalian (rat) sensory neurons the inactivation of the high-threshold current showed all the properties of the calcium-dependent inactivation already described (Veselovsky and Fedulova 1983; Fedulova *et al.* 1985). In mammalian neurons the 'non-specific' outward current is less expressed than in mollusc cells; therefore the decay of the high-threshold current here could be perfectly approximated (in a wide range of test potentials) by a single exponent with a time constant of several hundreds of milliseconds. The decay slowed down substantially in Ba^{2+}-containing solutions; in double-pulse measurements the amplitude of the test current demonstrated a U-shaped dependence on the amplitude of conditioning depolarization. The inactivation of the high-threshold current in these neurons also depended on the basal level of intracellular Ca^{2+}: its elevation from 10^{-9} to 10^{-6} M speeded up the inactivation substantially (Bossu *et al.* 1989*c*).

However, the low-threshold calcium current in the same cells inactivated itself in a completely different manner. The time constant of inactivation depended in this case only on the level of membrane potential, monotonically decreasing with depolarization. The difference is obvious from a comparison of results presented in Figs 4.4 and 4.5.

The voltage-dependent inactivation of low-threshold calcium channels has been clearly shown on nerve cells isolated from rat hypothalamus (Akaike *et al.* 1989). As can be seen from the example shown in Fig. 4.6, in this case the steady-state inactivation of low-threshold calcium currents was already practically complete at a holding potential of −70 mV; in other words, at the resting potential level when calcium channels are still closed and no Ca^{2+} ions can enter the cell.

The same difference in the mechanisms of inctivation of low- and high-threshold calcium channels has been found in other cells, even in oocytes. According to the observations of Fox and Krasne (1981), the membrane of Polychaeta oocytes revealed both types of calcium channels. The low-threshold calcium current inactivated rapidly (time constant of 10–50 ms) in a voltage-dependent way; the high-threshold current inactivated slowly (time constant of 300–600 ms), and the velocity of its inactivation depended not on the voltage but on the amplitude of the preceding calcium current.

Inactivation of low-threshold calcium channels has also been investigated

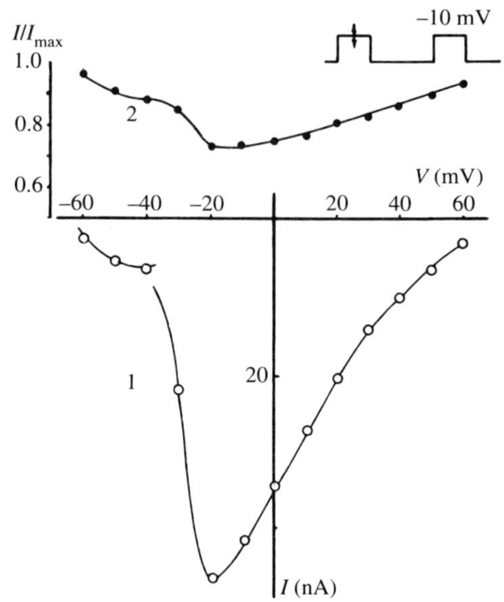

Fig. 4.4. Dependence of the maximal amplitude of the high-threshold calcium current in a rat dorsal root ganglion neuron on the current amplitude during conditioning depolarization (Fedulova *et al.* 1985). Curve (1) presents the *I–V* curve for the conditioning current, and curve (2) the normalized amplitude of the test current. The pulse protocol is shown in the inset.

in clonal malignant cells. As has been mentioned already, the membrane of non-differentiated neuroblastoma cells has predominantly low-threshold calcium channels; they inactivate in a purely voltage-dependent way. The same is true for mice myeloma cells, in which LTI calcium currents could be evoked by membrane depolarization to −50 mV (maximum current at −20 mV); inactivation developed at holding potential levels at which no influx of Ca^{2+} into the cell could take place (Fukushima and Hagiwara 1983). The fast-inactivating calcium current in the clonal pituitary malignant cells also inactivated in a voltage-dependent manner (Matteson and Armstrong 1984), whereas the dihydropyridine-sensitive high-threshold current inactivated in a calcium-dependent manner (Kalman *et al.* 1988).

Nevertheless, in some investigations this functional difference between the low- and high-threshold calcium channels has not been found to be so rigorous. According to the data of Bossu and Feltz (1986), the inactivation of low-threshold calcium channels in rat sensory neurons can also include a small calcium-dependent mechanism; on the contrary, Akaike *et al.* (1988) disclosed a small potential-dependent component in the inactivation of

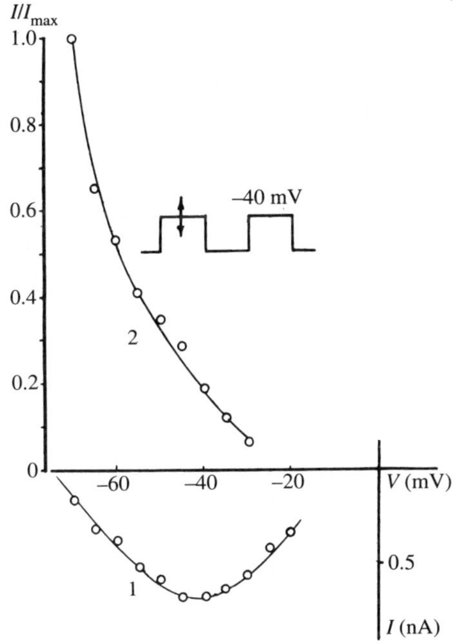

Fig. 4.5. Dependence of the maximal amplitude of the low-threshold calcium current on the current amplitude during conditioning depolarization (Fedulova *et al.* 1985). The figure is constructed in the same way as Fig. 4.4.

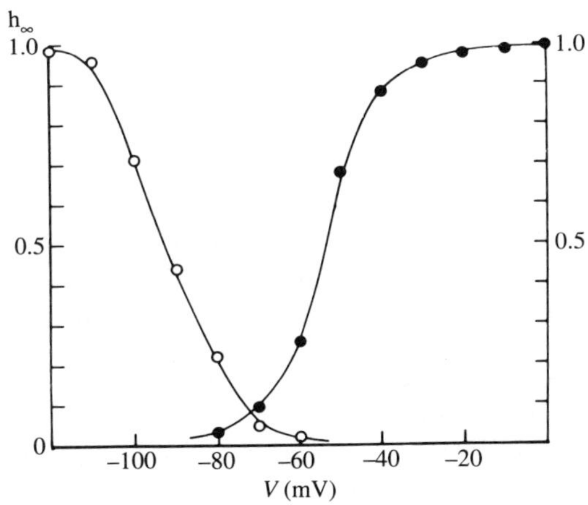

Fig. 4.6. Steady-state inactivation (open circles) and activation (closed circles) curves for the low-threshold calcium current in an isolated neuron from rat hypothalamus (Akaike *et al.* 1989). Inactivation was induced by a depolarizing potential shift to −30 mV preceding the test pulse by 1 s.

high-threshold calcium channels in frog sensory neurons. The same is probably true for high-threshold calcium channels in mammalian cardiomyocytes (Kass and Sanguinetti 1984; Lee *et al.* 1985).

The differences in the mechanisms of inactivation of low- and high-voltage activated calcium channels in neuronal tissue do not seem to be so clear for cells from some other tissues. In frog cardiomyocytes a U-shaped dependence of the test calcium current on the level of preceding depolarization has been shown in a double-pulse experiment; as already mentioned, this is typical for calcium-dependent inactivation (Mentrard *et al.* 1984). However, in frog skeletal muscle fibres inactivation of I_{Ca} has been observed at holding potentials at which calcium channels did not activate; in this case it was not of current-dependent but of voltage-dependent nature (Sanchez and Stefani 1983). In guinea-pig cardiomyocytes the inactivation of a slow inward current appeared to be independent of the replacement of Ca^{2+} by Ba^{2+} in the extracellular solution, which is typical for voltage-dependent inactivation (Cavalie *et al.* 1983).

Possible mechanisms of calcium-dependent inactivation

Although voltage-dependent inactivation plays some role in the functioning of calcium channels in nerve cells, the predominant mechanism of this process is still based on a recurrent action of Ca^{2+} ions which entered the cell from the external medium through the same channels. The presence of this unique mechanism of autoregulation may play a very important role in the function of the neuron as an integrated system (see Chapter 9 for details). Therefore, it would be very important to find out the intimate nature of this recurrent action.

As has been shown in Chapter 2, Ca^{2+} ions bind to the calcium channel during their passage through it. We have discussed the possibility that such binding might in some cases block the channel, but decided that it would be unrealistic from the physical point of view. Obviously, we have to start with the assumption that calcium-dependent inactivation is mediated by some intermediate steps, which interact with a kind of receptor structure at the inner mouth of the channel. This reaction induces some conformational changes in the channel, which convert it from the open (conducting) state into a non-conducting state. A formal description of such a mechanism has been given by Standen and Stanfield (1982). The model suggests the existence near the internal surface of the cell membrane of a special space into which Ca^{2+} ions passing through the calcium channel can enter directly. A similar idea has also been expressed by Chad and Eckert (1984*a*) who postulated the existence near the internal openings of the channels of special 'calcium domains' in which the concentration of free calcium must be elevated. Calcium-chelating substances introduced into the cell have only a limited access into this space;

this may be the reason why inactivation is never removed completely, even by introducing a large amount of chelators into the cell. The near membrane space exchanges Ca^{2+} ions with the rest of the cytoplasm and the intracellular depots; the ions present in it can also bind to a binding group at (or near) the calcium channel, inducing its inactivation. Choosing the constants of this model, the authors approximated the real time course of inactivation of the high-threshold calcium current. The K_d value for binding appeared to be about 7×10^{-7} M. This corresponds to the experimental findings of the same group (working on *Helix aspersa* neurons) that introduction into the cell of Ca-EGTA buffers either potentiated or depressed the peak value of I_{Ca} if the free calcium level was 1.8×10^{-7} M or 7.9×10^{-7} M, respectively (Plant *et al.* 1983). The natural level of free calcium in the same cells (measured with a calcium-selective electrode) was about 1.9×10^{-7} M. However, in our laboratory the K_d for calcium binding during inactivation was found to be about one order of magnitude lower (Doroshenko and Tsyndrenko 1978). It should be mentioned that such measurements may entail considerable errors, as has been shown by direct control of the intracellular level of free calcium by a calcium-selective microelectrode during cell perfusion with calcium-buffer-containing solutions (Byerly and Moody 1984). It emerged that it is difficult to predict exactly the level of intracellular free calcium from the characteristics of the Ca-EGTA buffer used. It was impossible to decrease $[Ca^{2+}]_{in}$ below approximately 10^{-7} M; the attempts to raise it above 10^{-6} M led to its spontaneous rise to the values of about 10^{-4} M (probably due to the activation of low-selective channels in the membrane and release of Ca^{2+} from intracellular stores, see Chapter 8). A depression of calcium currents was observed at $[Ca^{2+}]_{in} \cong 10^{-6}$ M (controlled by microelectrode measurements).

Until now there are no experimental indications about the nature of the suggested internal calcium-binding group. The chemical nature of this reaction can be predicted from the high-temperature dependence of the blocking effect of elevated $[Ca^{2+}]$. A drastic slow down of inactivation of calcium currents has been shown in cardiomyocytes after the intracellular introduction of proteases. This suggests that a special chemical group, destroyed by proteolysis, is involved in the inactivation of the corresponding calcium channels (Hescheler and Trautwein 1988).

One more feature of calcium channel functioning is very important for the understanding of the inactivation mechanism. The separation of calcium channels into two groups, one of which has voltage-dependent inactivation and the other calcium-dependent inactivation, coincides with another difference between these two groups. Calcium channels that demonstrate calcium-dependent inactivation are, at the same time, highly dependent on cellular nucleotide metabolism; calcium channels with voltage-dependent inactivation have no such dependence (Fedulova *et al.* 1985; Kalman *et al.* 1988). Injection of ATP into snail neurons promoted the restoration of inactivated high-threshold calcium currents, as did the injection of 10^{-4} M cyclic adenosine 3′,5′-monophosphate (cAMP); the effect was not observed on barium currents (Yatani *et al.* 1983). All these correlations prompted the suggestion that some of the intermediate links in the mechanism of the recurrent action of Ca^{2+} ions on calcium channels may be, at the same time, parts of nucleotide metabolic systems. This question will be discussed in more detail in Chapter 6.

A special problem is the presence in many neurons of a small component of voltage-operated calcium conductance which does not undergo inactivation, even during prolonged stimulation of the cell. This component has some special properties (for instance, it is less sensitive to the toxic action of penetrating Ba^{2+} ions). Most probably it is produced by a special subgroup of calcium channels. High-threshold non-inactivating (or very slowly inactivating) channels have already been described in the preceding chapter. A special group of very slow calcium channels has been found in guinea-pig cardiomyocytes (Lee *et al.* 1984).

Summary

Calcium channels, like other voltage-operated ion channels pass into a non-conducting (inactivated) state during long-lasting membrane depolarization. However, in most cases these channels inactivate in a special way, due to the recurrent action of calcium ions which entered the cell through them. The presence of this intracellular feedback system may be of major functional importance because it can regulate the influx of calcium depending on the metabolic state of the neuron. The precise mechanism of calcium-dependent inactivation is not yet clear; probably it includes some biochemical reactions as intermediate links. However, some calcium channels (predominantly the low-threshold ones) inactivate in a 'usual' potential-dependent way; these channels also have some other functional properties which are more characteristic of sodium or potassium channels.

5. Pharmacological characteristics and molecular structure of calcium channels

History

A characteristic feature of many types of ion channels is their ability to bind some natural substances very specifically. One of the best-known examples is the binding by sodium channels of tetrodotoxin (TTX) obtained from a Pacific fish, *Tetrodon*. The discovery of tetrodotoxin played an extremely important role in the development of modern ideas about ion channels. Some other specific substances have been found to block sodium channels; most of them are used by living organisms for defence or attack. The production of artificially modified natural toxins, including their radioactive labelling, became a very convenient instrument for the investigation of the functional structure of sodium channels, and determination of their location, density, etc.

Quite paradoxically, for a long time no natural substances that could be used in the same way for the study of calcium channels were detected. Apparently, the blocking of calcium channels is a much less effective way to kill the prey, and such a mechanism did not emerge in evolution. Only recently a specific natural blocker of calcium channels (but much less effective than TTX) has been isolated from some tropical predatory molluscs.

A more or less specific way to influence the activity of calcium channels is the use of di- and trivalent cations, which effectively interact with their internal binding group (see Chapter 2). However, it is hard to employ this method for pharmacological purposes because of the widespread side-effects of these ions. For this reason, artificially synthesized substances were mostly used for specific influences on calcium channels. The first success in this respect was achieved by Fleckenstein *et al.* (1969), who synthesized a group of phenylalkylamine derivatives, including verapamil. Even more important was the discovery of 1,4-dihydropyridines, which soon became one of the most widely used cardiac drugs, acting through the blockage of calcium channels. The synthesis of many modifications of these substances opened the way for their use as channel ligands in the chromatographic separation of channel-forming proteins from solubilized cellular membranes.

Phenylalkylamines

Verapamil and its methoxy-derivative, D-600, block calcium currents in many excitable cells, including neurons, at concentrations of about 10^{-5} M (Kohlhardt *et al.* 1972*b*; Baker *et al.* 1973*b*; Kostyuk and Krishtal 1977*a*). The structure of this substance is shown in Fig. 5.1.

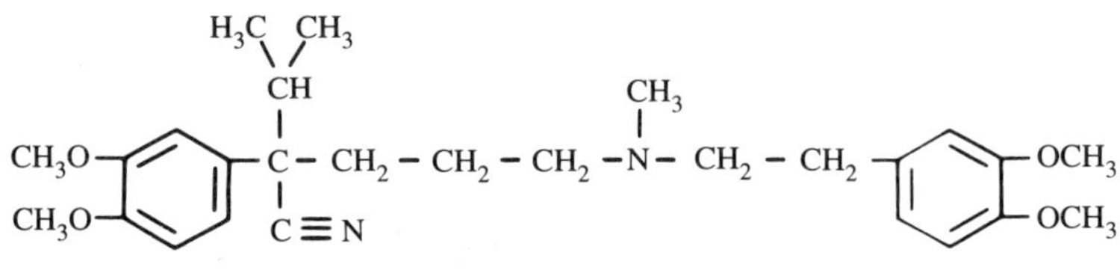

Fig. 5.1. The structure of verapamil.

Although the K_d of the blocking effect varies with the change in extracellular divalent ion concentration, this block is not due to competitive binding of the drug to the site in the channel that binds divalent cations. The location of the site of drug action is not quite clear. It is known from experiments on cardiac muscle fibres that the effect of phenylalkylamines increases during membrane depolarization and repetitive stimulation ('use-dependence', according to Pelzer *et al.* 1982; Trautwein *et al.* 1983). It has been shown in the same cells that the effect depends on the presence of a charge on the drug molecule. The pK_a value of the nitrogen at the compound D-600 is 8.5–8.6; this means that about 7 per cent of the molecules are not charged at normal pH. The compound D-890, on the contrary, has a permanently charged quarternary nitrogen. It was shown that D-890 does not affect calcium currents at 50 times higher concentrations than the effective dose of D-600. It has been concluded that the uncharged molecules of D-600 can penetrate through the lipid bilayer and reach the receptor site from inside the cell. The continuously charged D-890 cannot penetrate the membrane, but it inhibits calcium currents when injected directly into the cell (Hescheler *et al.* 1982).

However, in perfused snail neurons verapamil (and other organic calcium-channel antagonists) when injected inside the cell to block calcium currents were at least one order of magnitude less effective than when applied outside. This has been observed by Doroshenko in our laboratory and by Nishi *et al.* (1983). Therefore, at least for such cells it seems doubtful that intracellular penetration of verapamil is a prerequisite for its blocking action on calcium channels; it seems more probable that antagonists reach their binding sites from within the lipid phase of the membrane.

1,4-Dihydropyridines

The requirement to use phenylalkylamines in comparatively high concentrations indicates that the specificity of their action may not be very high. More specific are substances from the other group—the derivatives of 1,4-dihydropyridines (DHP). Figure 5.2 presents the structure of some of them. The high effectiveness of DHPs was first discovered on vascular smooth muscle fibres and then on cardiomyocytes. The depression of contractions was observed at nanomolar concentrations (Vater *et al.* 1972; Fleckenstein 1977); it developed in parallel with blockage of ^{45}Ca influx into the fibres (Cauvin *et al.* 1983). Contractions induced by membrane depolarization were depressed more effectively by DHPs than expected from the action of DHPs on voltage-operated calcium channels (Kazda *et al.* 1983). The use of labelled 1,4-dihydropyridines has shown their highly specific binding (K_d below 1 nM) to smooth and cardiac muscle membranes (Bellemann *et al.* 1981; Triggle and Janis 1984).

The investigation of the action of DHP on nervous tissue gave less consistent results. The binding of labelled DHPs to brain membranes seems to be of high affinity ($K_d \cong 1$ nM) (Ferry and Glossmann 1982; Bellemann *et al.* 1983). Marked binding of labelled nitrendipine in brain slices has been observed in regions rich in synaptic junctions (Murphy *et al.* 1982). In phaeochromocytoma cells a 50 per cent depression of ^{45}Ca influx has been observed at nitrendipine concentrations of several nanomoles (Toll 1982). However, direct measurements of calcium currents in snail neurons have shown their depression only at millimolar concentrations of DHPs (Kostyuk

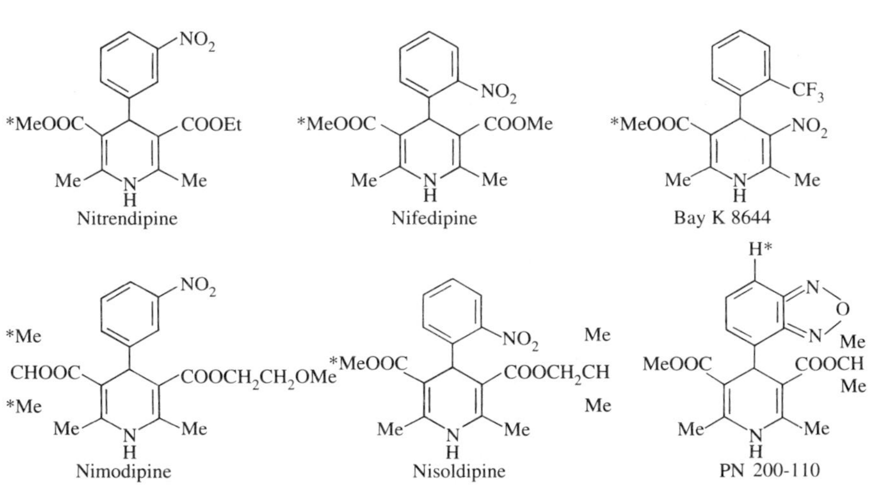

Fig. 5.2. The structures of different dihydropyridines.

a*et al.* 1983; Nishi *et al.* 1983). Calcium currents carried through different types of calcium channels in mammalian neurons were affected to a different extent; but even the sensitive ones could be blocked only by micromolar concentrations of antagonists (see below). Thus it seems that several types of DHP-receptors may exist in neuronal membranes, and high-affinity binding is not necessarily related to blocking of calcium channels.

The molecular mechanism of 1,4-dihydropyridine action can only be defined in general terms. The presence of the tertiary nitrogen in the pyridine ring is obligatory; its substitution by an NH-group removes the activity. Oxidation of the dihydropyridine ring has the same effect. Changes in the side-chains at positions 2, 3, 5, and 6 modulate the activity in a wide range. More optimal are the chains at positions 3 and 5 containing complex ether groups, and short alkyl chains in positions 2 and 6 (although the presence of an NH_2-group is also possible). At position 4 the presence of a substituted phenyl ring is most optimal; the type of substitution also has some importance. More details about the structural properties of DHPs can be found in the reviews of Janis and Triggle (1983, 1984), Kass and Arena (1989), and Triggle *et al.* (1989). Obviously, extensive stereochemical calculations are still necessary for the understanding of the role of these modifications in the determination of the calcium antagonist effect of DHPs.

Calcium ions play a certain role in the binding of DHPs. According to the data of several investigators, their removal (by chelators) disturbs the binding. This has been shown on cardiomyocytes by Glossmann *et al.* (1983*a*) and on brain slices by Gould *et al.* (1982). Binding can be restored by addition of Ca^{2+} and other divalent cations (in the sequence $Mn^{2+} > Ca^{2+} > Mg^{2+} > Sr^{2+}$). However, La^{3+} and Co^{2+} inhibited DHP binding. In this respect calcium channels look like metalloproteins in which divalent cations serve as a link between the ligand and the channel.

DHP-binding sites are definitely different from those for phenylalkylamines; they are different from binding sites for penetrating and blocking divalent cations because no real competitive relations exist between these agents (although at increased divalent ion concentrations the effectiveness of DHP binding can decrease somewhat). DHP-binding sites are also different from the site that regulates the ion selectivity of calcium channels, because the transformation of these channels into sodium-conducting channels cannot be affected by DHPs (Kostyuk *et al.* 1983).

A very interesting finding is that some structural modifications in the side-chains of the DHP molecule transform the latter from a blocker into an activator of calcium channels. The structure of one such compound (Bay K 8644) is shown in Fig. 5.2. Bay K 8644 potentiated the influx of Ca^{2+} ions and contraction in smooth muscle (Anderson *et al.* 1984), secretion of calcium-triggered noradrenalin release in phaeochromocytoma cells (Albus *et al.* 1984), and calcium inward currents in cardiomyocytes (Cohen and Chung 1984). The main effect of Bay K 8644 on single calcium channels consists in prolongation of their open state, leading to an increase in the integral calcium current (Fox *et al.* 1984; Hess *et al.* 1984). It should be stressed that antagonistic DHPs (nifedipine) are competitive inhibitors of the agonistic action of Bay K 8644 (Schramm *et al.* 1983); this means that their binding sites at the channel are, if not identical, then at least very similar.

Benzothiazepines

These synthetic substances form a special group of calcium-channel blockers. Diltiazem is the best-known representative (Fig. 5.3). Its L-form (but not the D-form) binds with high affinity to the microsomal fraction from skeletal muscles (K_d = 39 nM at 2°C and at 37 nM at 30°C); the binding depends very much on the simultaneous action of other antagonists—the DHPs inhibit it. On the contrary, application of L-diltiazem considerably increased the binding effectiveness of DHP. This may indicate the presence

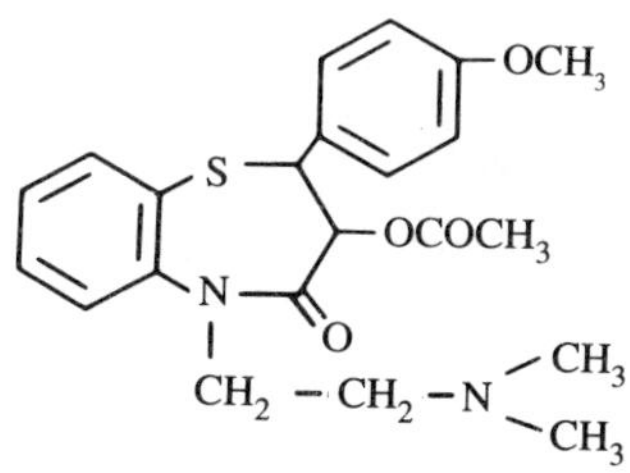

Fig. 5.3. The structure of diltiazem.

of allosteric interactions between the dihydropyridines and the benzothiazepine-binding sites (Ferry and Glossmann 1982; Glossmann *et al.* 1983*b*). The mechanism of such a positive allosteric effect is not clear; it has been suggested that diltiazem may transfer a 'silent' population of calcium channels into a state in which they can bind DHPs with high affinity (Ferry *et al.* 1983). Some data indicate that structural reorganizations may occur in the channel complex that binds DHPs; inactivation analysis has shown that after the action of diltiazem the molecular weight of the channel complex decreases, which may be due to dissociation of oligomers (Goll *et al.* 1984). Other substances, for instance phenylalkylamines and diphenylalkylamines, may also interact with the binding site for diltiazem; these exerted a negative but not positive allosteric effect on dihydropyridine binding (Murphy *et al.* 1983).

Calmodulin antagonists

It has been shown in several laboratories that many of the above organic calcium-channel blockers, including the 1,4-dihydropyridines, also interact effectively with the well-known intracellular calcium-binding protein, calmodulin, and inhibit its binding with Ca^{2+} ions. This effect does not seem to be due to their action directly on the calcium-binding site but rather to some conformational changes in the whole calmodulin molecule, altering its binding properties. Therefore it was natural to suggest that the calcium channels themselves have a calmodulin-like structure, which participates in their response to organic channel blockers. One of the ways to check this suggestion is to try the effect on calcium channels of substances known as calmodulin antagonists. In fact it has been shown that such antagonists (trifluoperazine, chlorpromazine, R 24571, and others) inhibited, in a competitive way, the binding of 1,4-dihydropyridines to rat brain membranes, with ED_{50} 1–20 μM (Thayer and Fairhurst 1983). They also selectively blocked the influx of Ca^{2+} ions into chromaffin cells (Clapham and Neher 1984), calcium action potentials and hormone secretion in malignant pituitary cells (Sand *et al.* 1983), and calcium currents and behavioural responses in Infusoria (Hennessey and Kung 1984; Ehrlich *et al.* 1988).

In our laboratory the effect of these substances has been tested on perfused snail neurons (Doroshenko *et al.* 1988*c*). The addition of chlorpromazine to the extracellular solution rapidly depressed calcium currents, with EC_{50} about 5×10^{-4} M. A similar effect could be exerted by trifluoperazine and R 24571; the last substance was even more effective (10^{-5} M). It has to be mentioned that chlorpromazine and trifluoperazine in lower concentrations (about 10^{-5} M) definitely potentiated the calcium currents, as can be seen from the dose–response curves shown in Fig. 5.4. Only the amplitude of the current was changed; the kinetic characteristics remained unaltered. Both effects were induced from the *external* side of the neuronal membrane; introduction of the substances into the neuron exerted, on the contrary, a stabilizing effect on calcium currents, which will be analysed in more detail in Chapter 6.

The existence of a calmodulin-like structure related to the calcium channel has been suggested for the muscle fibre membrane (Johnson *et al.* 1983; Johnson 1984). On the basis of this suggestion an attempt has been made to explain the action of organic antagonists, as well as intracellular Ca^{2+} ions, on calcium channels. The authors postulated that binding of Ca^{2+} ions to this structure (which may be located near the inner mouth of the channel) induces conformational changes in the channel and inactivates it. Organic antagonists may increase the affinity of this structure to Ca^{2+} ions in such a way that Ca^{2+} ions start binding to it at lower intracellular concentrations.

However, this suggestion seems to be incorrect, at least for the neuronal

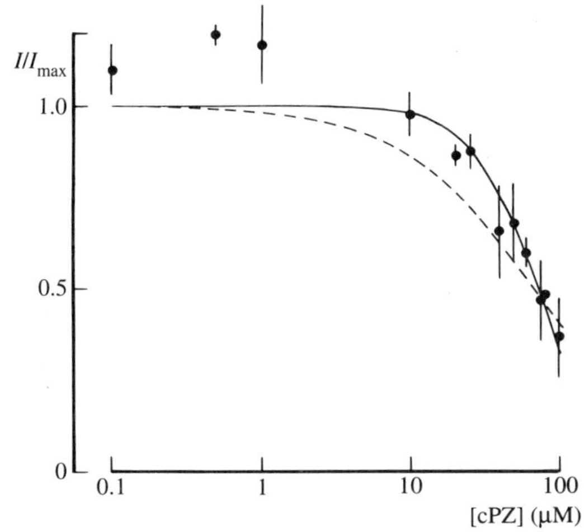

Fig. 5.4. Concentration dependence of the effect of an external application of the calmodulin inhibitor chlorpromazine (CPZ) on the calcium current in a snail neuron (Doroshenko *et al.* 1988*c*). Points indicate the normalized values of peak current amplitude; the solid line corresponds to bimolecular binding, and the dotted line to monomolecular binding, of the drug to the channel.

membrane. Calmodulin antagonists exerted their blocking effect in this case only from the external side of the membrane, and the calcium-dependent inactivation most probably has a different mechanism (see Chapter 6).

Quite recently an attempt was made to define the possible calmodulin-like structure in the calcium channel directly, by comparing the recently described primary structure of the α_1-subunit of the calcium channel (see below) with the already known primary structure of calmodulin. A segment has been identified in the region of residues 961–987 of the subunit polypeptide which has 42.5 per cent of amino acids in common with the 76–125 region of calmodulin (Semin and Ivanov 1989). Further analysis is necessary to discover the possible functional meaning of this region.

Natural agonists and antagonists

As already mentioned, natural toxins that specifically affect calcium channels have been detected recently. One of them is maitotoxin, isolated from Flagellata and investigated by Miyamoto *et al.* (1984). It substantially prolonged the duration of the calcium component of the action potential in insect muscles at concentrations of 5×10^{-9}–10^{-8} g l^{-1}. The effect of maitotoxin could be inhibited by inorganic calcium-channel blockers (Co^{2+}).

The mechanism of the activating effect of this toxin on calcium channels has to be analysed in more detail, especially by direct recording of calcium currents.

Another toxin that evoked great interest is the ω-conotoxin isolated from the tropical fish-hunting molluscs *Conus geographicus* (toxin M VII A) and *Conus magnus* (toxin M VII B). The amino-acid sequences of these toxins are somewhat different; the first was synthesized artificially under the name ω-CgTX and became available for experimental use (Cruz *et al.* 1987; McClesky *et al.* 1987; Olivera *et al.* 1987; Suzuki and Yoshioka 1987). Both polypeptides compete for the same binding site on the neuronal membranes and their action seems to be based on a similar mechanism.

A very important property is the specificity of ω-CgTX towards brain tissue; it does not affect calcium currents in cardiac and skeletal muscle fibres. Therefore it might be a convenient instrument for the chromatographic separation of proteins forming high-threshold calcium channels in neuronal structures.

Pharmacological properties of different types of calcium channels

In analysing the pharmacological properties of calcium channels, it is very important to take into account their non-homogeneity. The general rule is that the described agonists and antagonists act predominantly on high-threshold channels; their effect on the low-threshold ones is very weak or even absent. According to the data from many laboratories, the high-threshold non-inactivating calcium channels in the membrane of dorsal root ganglion neurons can be blocked by phenylalkylamines and 1,4-dihydropyridines (in concentrations of 10^{-5}–10^{-4} M) without substantial change in the activity of LTI and HTI channels (Boll and Lux 1985; Nowycky *et al.* 1985*a*; Fox *et al.* 1987*a*; Kostyuk *et al.* 1988*a*). Bay K 8644 somewhat increased the activity of high-threshold non-inactivating channels (due to prolongation of their mean open time) but was without any effect on low-threshold channels. The same is true for clonal pituitary cells (Simashko *et al.* 1988; Cohen and McCarthy 1987). According to the suggestion of Nowycky *et al.* (1985*b*), this calcium channel agonist transfers the gating mechanism of the latter in a special functional mode characterized by the appearance of very long open states. However, according to our measurements, Bay K 8644 only shifts the equilibrium in favour of one of the open states which are also present in the kinetics of this type of calcium channel in the absence of the drug; a definite prolongation is observed of just the 'fast' component in the distribution of the open times, without any detectable change in the 'slow' component (Kostyuk *et al.* 1988*a*). Figure 5.5 demonstrates examples of the described effect of Bay K 8644 on the activity of single HTN channels in a mouse sensory neuron.

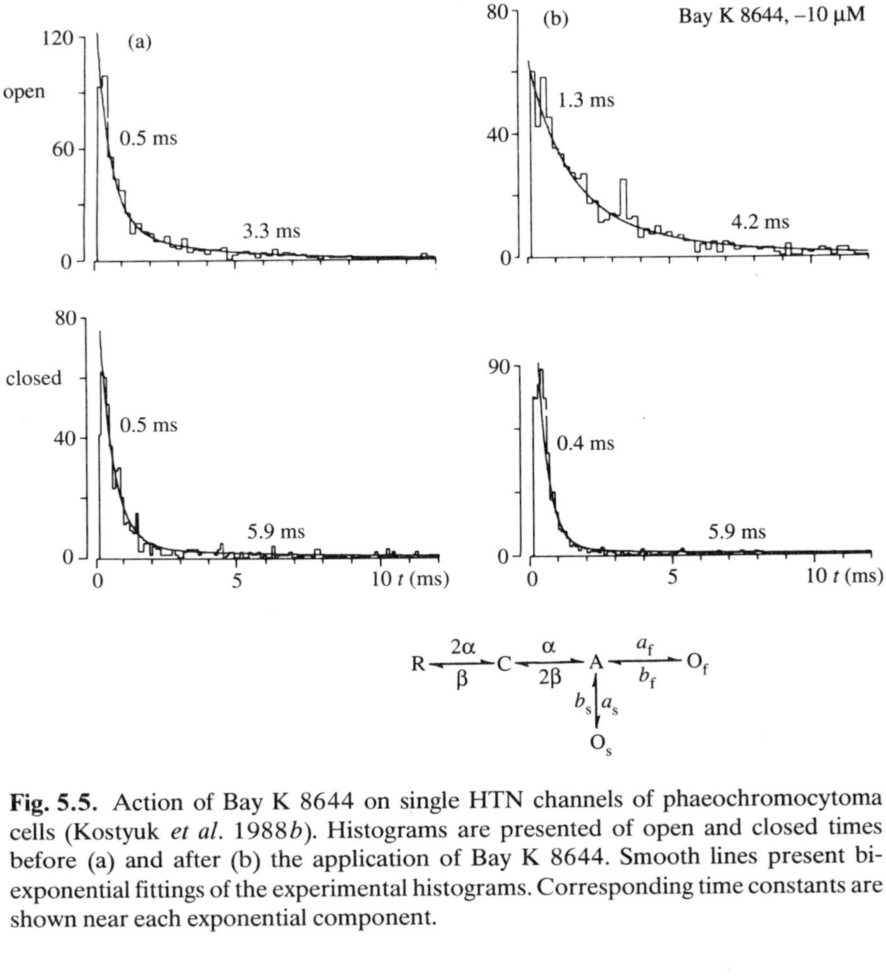

Fig. 5.5. Action of Bay K 8644 on single HTN channels of phaeochromocytoma cells (Kostyuk *et al.* 1988*b*). Histograms are presented of open and closed times before (a) and after (b) the application of Bay K 8644. Smooth lines present bi-exponential fittings of the experimental histograms. Corresponding time constants are shown near each exponential component.

In the membranes of other excitable cells, the difference in the action of the discussed substances on low- and high-threshold calcium channels is expressed even more because of much higher sensitivity of the high-threshold calcium channels to these drugs: in smooth, cardiac, and skeletal muscle fibres it is by several orders higher than in nerve cells (Bean 1985; Kongsamut *et al.* 1985; Brown *et al.* 1986; Cognard *et al.* 1986). In fibroblasts the low-threshold calcium channels are also completely insensitive to DHPs and verapamil (Peres *et al.* 1988).

Thus, the differences in pharmacological sensitivity may be used as an important additional criterion for the separation of different types of voltage-operated calcium channels. However, this conclusion cannot be automatically extended to all tissues. For instance, in the recent investigation of nerve cells isolated from rat hypothalamus slices (performed with the use of a new technique of rapid application of substances) it has been shown that organic calcium-channel antagonists inhibit low-threshold channels at about the same concentrations as they inhibit the high-threshold channels in sensory neurons (Akaike *et al.* 1989). The corresponding dose–response curves are presented in Fig. 5.6. It is interesting that in this case Bay K 8644 exerted not an agonistic but a weak antagonistic effect on LTI channels, with K_d about 10^{-5} M. A very effective blocker of LTI channels in this case was flunarizine, a substance from a new group (diphenylpiperazine- and difluoro-derivative of cinnarazine) which has been known before only by a slow developing inhibition of calcium currents in smooth muscles. A predominant blocking effect of nifedipine on low-threshold channels has also been shown recently in cultured neurons from rabbit brain cortex (Dichter and Zona 1989).

Quite specific blockers of high-threshold calcium channels are the ω-conotoxins. ω-CgTX blocked high-threshold calcium currents at concentrations of about 50 pM, but hardly affected low-threshold currents (some response appeared only at concentrations of about 1 μM). But this toxin cannot be considered as a universal tool for the identification of these channels because it did not affect them in muscle fibres to the same degree as in motor nerve terminals (Anderson and Harvey 1987), although recently the blocking effect of ω-CgTX has been demonstrated on cholinergic nerve terminals in ileum; perhaps the calcium channels here are of a special type (Landy and Frew 1988). Substantial quantitative differences have been observed in the action of this toxin in different animal species; it was more potent in frog and chick than in rat (Suszkiw *et al.* 1987).

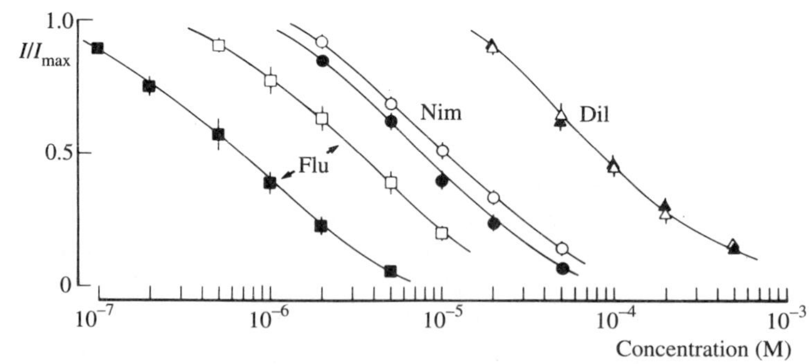

Fig. 5.6. Blockade of low- and high-threshold calcium currents in rat hypothalamus neurons (Akaike *et al.* 1989). Closed symbols, low-threshold currents; open symbols, high-threshold currents. Flunarizine (Flu), nimodipine (Nim), and diltiazem (Dil) were tested. Currents were evoked by voltage steps from −100 mV to −30 mV and from −60 mV to 0 mV, respectively.

The use of pharmacological ligands for the isolation of calcium channels

The detection of substances that specifically affect only the calcium channels opened the way for attempts to isolate the channel-forming proteins by affinity chromatography. This has mostly been done with the help of DHP derivatives which irreversibly bind to calcium channels.

In the first experiments carried out mainly on skeletal muscle membranes, it was revealed that the isolated proteins consist of several subunits, one of which (α) is quite heavy (135–170 kDa) and the others (β, γ) lighter (30–50 kDa) (Ferry *et al.* 1983; Curtis and Catterall 1984; Borsotto *et al.* 1985; Barhanin *et al.* 1987). The α- and β-subunits could be phosphorylated by cAMP-dependent protein kinase, like the natural calcium channels in muscle fibre membranes (Curtis and Catterall 1984; Hosey *et al.* 1986). However, attempts to clone the corresponding mRNA, and to separate and express the genes of these subunits have met some difficulties. These were due to the fact that the heavy subunit consists of two subunits (α_1 and α_2) with very similar molecular weights (175 and 143 kDa) from which only one (α_1) formed functioning calcium channels when incorporated into a phospholipid bilayer (Perez-Reyes *et al.* 1989). It could be phosphorylated by cAMP-dependent protein kinase and could bind 1,4-dihydropyridines. Monoclonal antibodies to this subunit inhibited dihydropyridine-sensitive calcium currents in muscle fibres. The α_2-subunit, which is heavily glycosylated, did not form calcium channels in a bilayer (Imagawa *et al.* 1987*a*; Burgess and Norman 1988; Campbell *et al.* 1988; Froehner 1988); however, it participates in some way in DHP binding because the removal of its sialic acid terminals decreased such binding by about 50 per cent. The relations of all the subunits in the muscle membrane calcium channel can be seen in a scheme proposed by Takahashi *et al.* (1987) and illustrated in Fig. 5.7.

The primary structure of the α_1-subunit has recently been reported (Mikami *et al.* 1989). It shows a fundamental similarity to the structure of voltage-operated sodium channels; both include four homologous domains, each of which contains several transmembrane spanning regions. This is discussed in detail in a review by Catterall (1989). Unlike the α_1-subunit, the β-subunit has no homology to other membrane proteins (Ruth *et al.* 1989).

Data about the molecular structure of calcium channels from neuronal membranes are less definite. Affinity columns with DHPs are not very convenient in this case because of the low effectiveness of their binding. For this reason affinity columns with ω-CgTX were used recently for the isolation of calcium channels from brain homogenates. This toxin, as already mentioned, is a specific blocker of high-threshold neuronal calcium channels, it does not affect the low-threshold ones. A polypeptide with molecular weight (MW) 210–220 kDa has been isolated, together with a

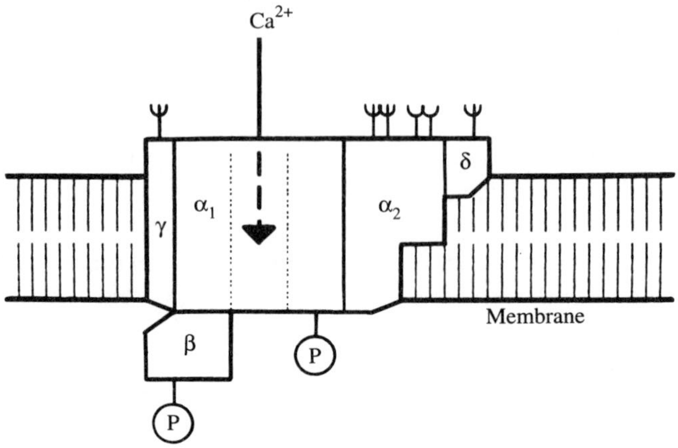

Fig. 5.7. The suggested composition of subunits in the voltage-operated high-threshold calcium channel. The α_1-subunit contains the DHP-binding site. The phosphorylation sites are indicated by (P).

component of 170 kDa. Possibly, the latter corresponds to the α_1-subunit of the calcium channels isolated from muscle membranes (Barhanin *et al.* 1988). A similar polypeptide (MW 222 kDa) has been isolated from the synaptosomal fraction of cultured brain neurons (Marqueze *et al.* 1988).*

Recently, a peptide with MW 85–95 kDa has been isolated from the *cytosolic* fraction of the skeletal muscle homogenate, which binds DHPs with high affinity (Krizanova *et al.* 1989). Antibodies to calcium channels inhibited this binding. This peptide is likely to represent a soluble precursor of the α_1-subunit, which can be incorporated in the cellular membrane and there form functioning calcium channels. A similar soluble precursor of TTX-sensitive sodium channels had been detected earlier in the cytosolic fraction from the brain homogenate (Lishko *et al.* 1987).

Conclusion

The main instruments for the analysis of structure and function of calcium channels, in contrast to the sodium ones, are several groups of synthetic organic substances (phenylalkylamines, 1,4-dihydropyridines, benzothiazepines, calmodulin antagonists). The substances from different groups may well bind to different binding sites in calcium channels, and between them both positive and negative allosteric interactions may exist. Some of these

* Primary structure and functional expression from complementary DNA of a brain calcium channel have just been obtained by Mozi *et al.* (1991). *Nature*, **350**, 398–402.

substances can act as channel agonists, affecting the gating mechanism of calcium channels. The effectiveness of action varies to a large extent for different types of channels and for channels from different tissues. 1,4-Dihydropyridines are most effective in smooth and cardiac muscles and much less efffective in brain tissue (although their specific binding in the latter case has the same high affinity as in muscles). A natural blocker (ω-conotoxin) has been found only for calcium channels in the neuronal membranes.

The substances have been used for the isolation by high-affinity chromatography of the proteins that form calcium channels. In the muscle membranes they consist of several subunits one of which (α_1) forms the channel in a phospholipid bilayer and binds the ligands with high affinity; other subunits obviously play a regulatory role. In neuronal membranes the calcium channels seem to include a similar subunit, however, this matter is not quite settled yet.

6. Modulation of calcium-channel functioning

History

The 'classical' concept concerning the functioning of ion channels considers them as being operationally independent of cellular metabolism. The genetic apparatus of the cell controls the synthesis of protein macromolecules, which are transported by the mechanism of intracellular transport to the relevant sites, incorporated into the surface membrane, and then act there for a while, being replaced in time by new ones. Metabolic processes create only the necessary prerequisite for the function of the channels: they form, through the activity of the metabolically driven pumps, constant transmembrane gradients of the main inorganic ions. As long as these gradients are present, the membrane maintains its ability to respond to external stimuli by active responses in the form of ionic currents triggered by conformational changes in channels that open the pathway for ions. Cellular metabolism is not necessary for these changes; the membrane has only to maintain its labile structure, characteristic of the living cell. This concept seemed to be quite universal; however, in the past few years it has become more and more obvious that it considerably simplifies reality and that channel function can, in fact, be under the direct control of intracellular metabolic processes; this is especially true for calcium channels.

The first indications of such metabolic modulation of calcium-channel function were obtained on cardiac muscle fibres. It is well known that the calcium-dependent plateau of the action potential in this cell is prolonged by the action of catecholamines; this prolongation is of especial importance during sympathetic augmentation of cardiac activity. Direct introduction of cyclic adenosine 3′,5′-monophosphate (cAMP) into cardiomyocytes produces a similar effect (Tsien *et al.* 1982; Vogel and Sperelakis 1981). The extracellular application of cAMP analogue (dibutyryl- and monobutyryl-cAMP), which can penetrate through the cell membrane and exert an effect similar to that of intracellular cAMP, also prolonged the duration of the plateau and increased the slow inward current, just like the catecholamines (Reuter 1974, 1979). On the basis of findings about the functional role of cyclic nucleotide metabolism in the activation of protein phosphorylation (Greengard 1978), a suggestion has been made that the catecholamine-induced potentiation of the calcium component of the cardiac action potential is mediated through the increased synthesis of cAMP by adenylate cyclase, and subsequent phosphorylation of the proteins responsible for calcium conductance by the cAMP-dependent protein kinase. Special investigations have shown that during such action no changes in selectivity or kinetic properties of the calcium current occur; consequently, the suggested phosphorylation increases the number of functioning channels and/or the duration of their opening (Reuter and Scholz 1977*b*). A recent analysis of the role of cAMP in the functioning of single cardiac calcium channels has shown that a prolongation of their open state is the main reason for the augmentation of calcium current (Reuter *et al.* 1982; Reuter 1987).

A possible role of cAMP in the modulation of calcium conductance in the neuronal membrane was suggested by Shimahara and Tauc (1977) and Klein and Kandel (1978). They recorded in *Aplysia* the synaptic effect exerted by a certain neuron on another one (or recorded directly the slow inward current in that neuron) and observed their potentiation under external application of serotonin (serotonin is a common neurotransmitter in mollusc ganglia). The same effect could be reproduced by cAMP injection into the cell, or by incubation of the ganglion in a solution containing phosphodiesterase inhibitors (phosphodiesterase destroys intracellular cAMP). As the inward flux of calcium is an obligatory link in synaptic action, it was concluded that serotonin increases such influx through an increase in intracellular cAMP level. Of course, in this case the change in synaptic action could be connected not directly with an increase of inward calcium current but with a decrease of the parallel outward potassium one; indeed, this idea has been suggested in further experiments of Klein and Kandel (1980). Nevertheless, the original findings stimulated more detailed investigations of the possible metabolic modulation of calcium-channel function, using direct recordings of calcium currents.

cAMP-dependent modulation of calcium currents in the neuronal membrane

The high-threshold calcium channels in the neuronal membrane have an important property that makes them very convenient for a study of the metabolic dependence of their function. During intracellular dialysis the corresponding calcium currents rapidly decreased in amplitude: in large mollusc neurons this occurred in several tens of minutes (Kostyuk and Krishtal 1977*a*; Byerly and Hagiwara 1982); in smaller mammalian neurons it occurred in a few minutes (Kostyuk *et al.* 1981*e*). The sodium and potassium currents remained almost unchanged during dialysis. The kinetic characteristics of the calcium currents during such a 'wash-out' did not change; its velocity was highly temperature-dependent. Examples of such dependence are shown in Fig. 6.1.

Obviously, a cytoplasmic factor, which can be easily washed out from the

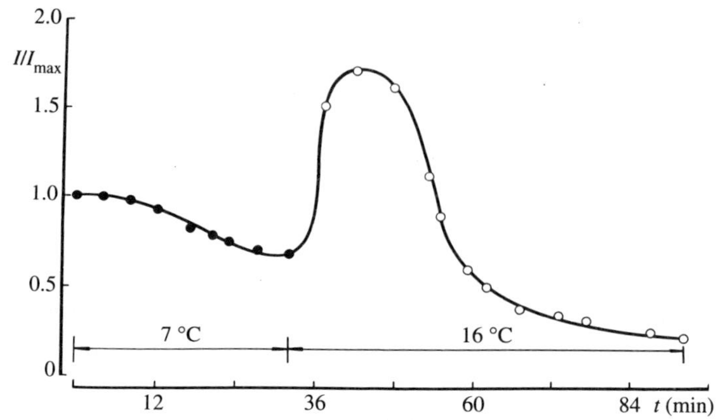

Fig. 6.1. Dependence of the 'wash-out' of calcium current in the course of perfusion of a snail neuron on temperature (Doroshenko *et al.* 1984*a*). The current amplitude at the beginning of perfusion in this and several subsequent figures is taken as unity.

cell or destroyed during its dialysis, is necessary for normal functioning of calcium channels. Introduction of different drugs in the dialysing solution can be used in the search for such a substance. Both in mollusc and mammalian neurons the introduction of cAMP together with ATP and Mg^{2+} (the necessary co-factor in ATP hydrolysis) in many cells not only prevented a further decrease of calcium currents but sometimes restored them to the initial level. Separate introduction of each of these substances exerted only a weak stabilizing effect (Fedulova *et al.* 1981; Doroshenko *et al.* 1982). In snail neurons a maximal effect was observed at a cAMP concentration of about 10^{-4} M, although some restoration could be seen even at micromolar concentrations. Optimal concentrations were 2 mM ATP and 3 mM Mg^{2+}. After reaching a maximal effect, the amplitude of the calcium currents started to decrease again but less rapidly than during the dialysis with simple saline solution. The introduction of cGMP did not exert any effect on the 'wash-out' of calcium currents. Figure 6.2 illustrates the restoration of calcium currents obtained in such a way in a rat sensory neuron, and Fig. 6.3 shows a similar effect on a snail neuron. In Fig. 6.3 the dependence of the effect on the cAMP concentration can be seen; the requirement for the combined introduction of cAMP and ATP is also obvious. It should be kept in mind that, during such experiments, cell dialysis with saline solutions results in a considerable increase in the intracellular level of free calcium, unless special measures are taken; as already mentioned in Chapter 4, such an increase effectively depresses the calcium conductance. However, specific measurements of free calcium level in the dialysing solution have shown that the restoring effect of cAMP is not connected with this level.

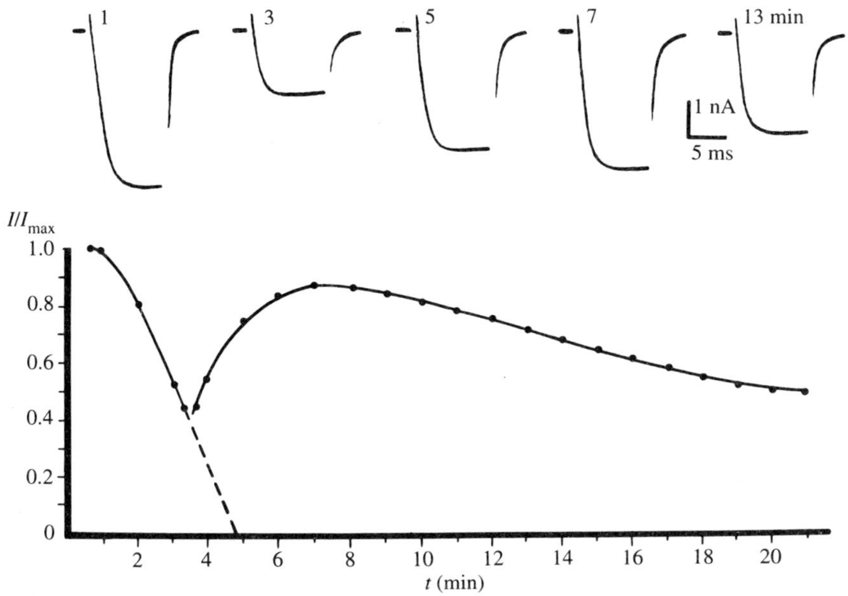

Fig. 6.2. Time-dependence of the 'wash-out' of calcium current in perfused rat dorsal root ganglion neuron before and after intracellular application of cAMP + ATP + Mg^{2+} (Fedulova *et al.* 1981). The arrow shows the moment of application. In the inset, examples of currents recorded at different times after the beginning of the perfusion are shown.

These experiments indicate that the activity of membrane-bound enzymes is retained in the conditions of intracellular dialysis or perfusion and that they can be activated by introduction into the cell of the corresponding substrates. This is supported by a series of other experiments. Thus, addition of fluoride ions to the perfusate in concentrations that activate membrane adenylate cyclase (several mM) together with ATP and Mg^{2+} also restored calcium currents. On the contrary, addition of Cu^{2+} (adenylate cyclase inhibitors) speeded up the 'wash-out'. These results can be interpreted in such a way that in the presence of sufficient ATP in the cell the fluoride-activated adenylate cyclase can sufficiently increase the near-membrane concentration of cAMP to keep calcium channels functioning. Large amounts of fluoride, as already mentioned, exerted an opposite effect: they rapidly and irreversibly switched off the calcium conductance. This effect also seems to be metabolically mediated: it slowed down considerably as the temperature was lowered. A small but stable potentiation of high-threshold calcium currents in some snail neurons has also been observed after activation of adenylate cyclase by forskolin (Doroshenko *et al.* 1988*b*).

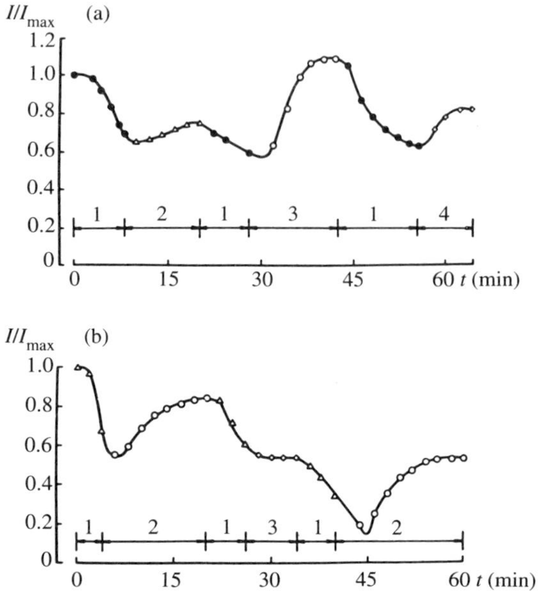

Fig. 6.3. Dependence of the rate and degree of restoration of the 'washed-out' calcium current on cAMP concentration and the presence of ATP on a snail neuron (Doroshenko *et al.* 1982). In (a) the perfusate contained 130 mM Tris-aspartate (1), 125 mM Tris-aspartate + 2 mM ATP + 3 mM Mg^{2+} (2–4). The cAMP concentration was 5×10^{-6} M (2), 5×10^{-4} M (3), 5×10^{-5} M. In (b) the initial solution (1) was 130 mM Tris-aspartate, then cAMP (5×10^{-5} M) was introduced with 2 mM ATP (2) and without it (3).

This mechanism of cytoplasmic control of membrane calcium conductance is characteristic only for the high-threshold channels. The low-threshold calcium channels, as already mentioned, are very resistant to alterations of intracellular processes; they can retain their function in isolated membrane patches for a long time. Rapid inactivation of low-threshold channels is not related to the action of intracellular calcium ions (Carbone and Lux 1984; Fedulova *et al.* 1985). Intracellular introduction of fluoride or cAMP also did not modulate the activity of low-threshold channels (Dupon 1986; Carbone and Lux 1987*a*). In sheep pituitary gonadotrophs injection of cAMP supported the function of high-threshold but not low-threshold calcium channels (Mason and Sikdar 1989).

It should be mentioned that in experiments when cAMP is injected from a microelectrode (by iontophoresis or under pressure) into intact snail neurons an opposite effect has often been observed: a reversible depression of calcium currents (Kononenko *et al* 1983). Probably, the mechanism of cAMP action in this case was complicated by the liberation of calcium from intracellular stores or an increase in the intracellular pH. In dialysed cells the level of free calcium is largely stabilized, and direct effects of cAMP become more obvious. In similar experiments with pressure injections of cAMP in larger *Aplysia* neurons Pellmar (1981) observed, on the contrary, an induction of calcium inward current. A different effect has been observed in sensory neurons: cAMP injection selectively reduced the high-threshold inactivating (N) component of the calcium current by *speeding-up its inactivation* (Gross and MacDonald 1989). However, Kalman *et al.* (1988) on GH_3 cells found an opposite influence of injected cAMP or application of forskolin: a slow-down of inactivation of high-threshold calcium channels.

The experimental data described still do not prove directly the suggestion that the increase in intracellular cAMP level effects the calcium conductance through activation of the cAMP-dependent protein kinase A (PK-A) which, in turn, phosphorylates some proteins important for the functioning of calcium channels—although such an hypothesis seems to be most reasonable. More direct evidence has been obtained in experiments with the introduction in the dialysed neuron of the catalytic subunit of PK-A. PK-A in its inactive form is a complex of two regulatory and two catalytic subunits which, under the action of cAMP, dissociate into separate subunits; phosphotransferase activity is the function of the free catalytic subunits (see Glass and Krebs 1980). The active catalytic subunit can be obtained biochemically in pure form; its property seems to be identical for different tissues of different animals. On this basis, experiments have been carried out to study the effects of PK-A catalytic subunit obtained from rabbit myocardium introduced into dialysed snail neurons (Doroshenko *et al.* 1984*b*). As can be seen from the examples shown in Fig. 6.4, such an introduction stopped the 'wash-out' of high-threshold calcium currents and restored them, sometimes up to the initial level. The presence of ATP was necessary for the effect, the addition of cAMP was not. The stabilized calcium currents of constant amplitude could be recorded during long-lasting (several hours) cell perfusion. Removal of ATP from the cell resulted in rapid deterioration of the currents (see Fig. 6.4b); its readmission could partially restore the latter. The slowness of the observed changes could be attributed to slow diffusion of the large catalytic subunit molecules to their site of action. An opposite effect, depression of the PK-A, could be achieved by the use of specific inhibitors, for instance tolbutamide. It has been shown that on snail neurons tolbutamide exerts an effect on high-threshold calcium current similar to that of cAMP depletion (Doroshenko *et al.* 1984*b*; Fig. 6.5).

These experimental data support the concept that the modulation of calcium conductance in the neuronal membrane can occur by the phosphorylating activity of PK-A. In *Aplysia* neurons the injection (through a microelectrode) of the PK-A catylytic subunit purified from bovine myocardium also facilitated the generation of 'calcium' action potentials (Kaczmarek *et al.*

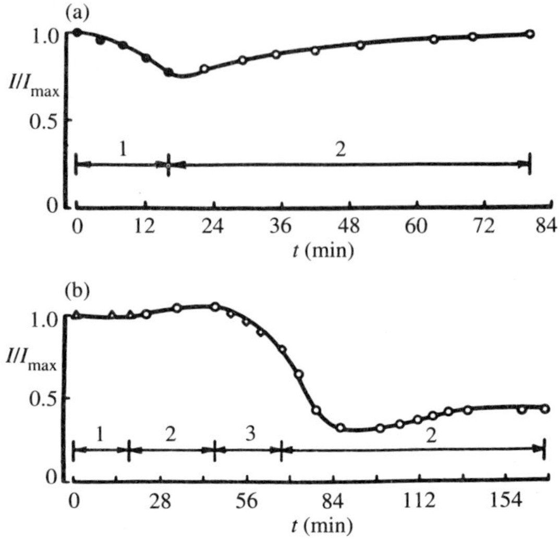

Fig. 6.4. The effect of intracellular administration of the catalytic subunit (CS) of cAMP-dependent protein kinase on the 'wash-out' of calcium current in a snail neuron (Doroshenko *et al.* 1984*b*). In (a) perfusion was performed with 125 mM Tris-aspartate + 2 mM ATP + 3 mM $MgCl_2$ (1) and the same solution containing 0.7 μM of CS (2). In (b) the procedure was repeated in the presence (2) and absence (3) of ATP.

1980). Chad and Eckert (1984*b*) have observed the restoration of decreased calcium conductance by intracellular injection of the PK-A catalytic subunit together with ATP and Mg^{2+}, or by the action of forskolin. ATP could not be replaced by its non-hydrolysable analogue (AMP-PCP). Barium ions rapidly abolished the supporting effect of phosphorylation. The authors concluded that the decrease in calcium currents during cell perfusion is due to progressive dephosphorylation of the channel-forming proteins; constant phosphorylation of the latter is necessary for normal activation of calcium channels. Natural inactivation of calcium channels may also be connected with channel dephosphorylation, as it slows down after the injection of cAMP or catalytic subunit (see also Armstrong and Eckert 1985, 1987; Chad and Eckert 1986; Eckert *et al.* 1986; Armstrong and Kalman 1988). cAMP supports the function of calcium channels expressed in *Xenopus* oocytes after injection of poly(A) mRNA from rat brain (Kaneko and Nomura 1987), and that of calcium channels in pancreatic B-cells (Henquin and Meissner 1983).

The similarity between the effects of an increase in intracellular calcium and depression of protein phosphorylation led to the suggestion that the blocking effect of calcium on calcium channels is also mediated through a

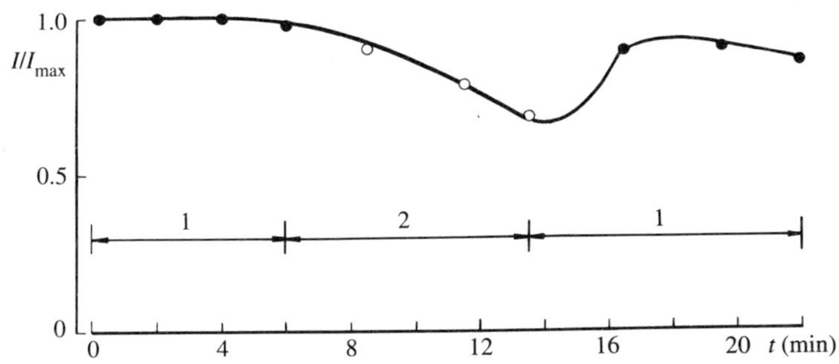

Fig. 6.5. Tolbutamide depression of the calcium current during cell perfusion of a snail neuron (Doroshenko *et al.* 1984*b*). The cell was perfused with 130 mM Tris-aspartate + 2 mM theophylline (1) and the same solution with 2 mM tobutamide added (2).

metabolic link. Potentiation of the dephosphorylation of the channel-forming protein phosphatase 2B (calcineurin) may play the main role in this process; this enzyme activates phosphatase 1 by dephosphorylation of an endogenous inhibitory peptide (Armstrong 1989) and phosphatase 1 then dephosphorylates the channel-forming protein.

This suggestion has been supported by work in several laboratories on different tissues. The action of a changed level of intracellular calcium on the effects of cAMP on calcium currents has been studied in snail neurons (Doroshenko *et al.* 1982). Figure 6.6 shows an example of the results. As can be seen, the introduction of cAMP together with ATP and Mg^{2+} into a cell perfused with a solution containing a very high level of Ca^{2+} (1 mM) removed the blocking effect of calcium. Direct determination of free calcium in the perfusing solution with a calcium-sensitive electrode has shown that this was not due to calcium binding by ATP or cAMP. Opposite changes in intracellular calcium caused by addition of calcium chelators (EGTA) produced an opposite effect: they potentiated the restoring effect of cAMP on calcium currents; an example is shown in Fig. 6.7.

One of the points of interaction between Ca^{2+} and cyclic nucleotides in their recurrent action on membrane channels could be the system of cellular phosphodiesterases (PhDE). The activity of PhDE is highly dependent on Ca^{2+} ions which activate it in micromolar concentrations through the formation of a complex with calmodulin (see Rasmussen *et al.* 1979). An increase in the intracellular calcium level will lead, through this mechanism, to a decrease of the cAMP level and correspondingly to the transition of calcium channels into an inactive state; on the contrary, at low calcium levels the activity of PhDE will be substantially depressed. The correctness of these

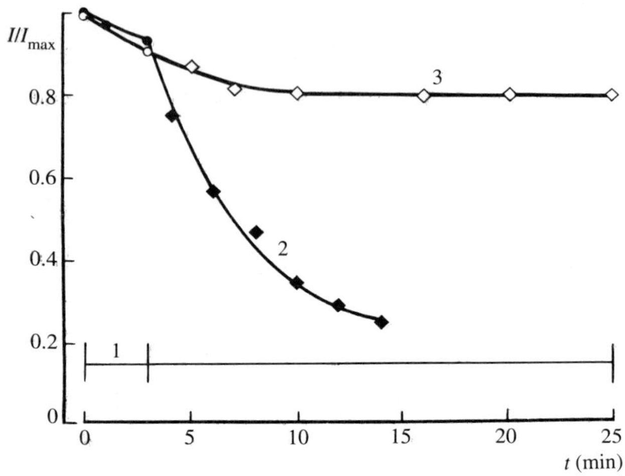

Fig. 6.6. Elimination of the blocking action on calcium current of intracellular Ca^{2+} with the addition of exogenous cAMP in a snail neuron (Doroshenko *et al.* 1982). The cell was first perfused with basic solution (1) to which 5×10^{-4} M $CaCl_2$ was later added (curve 2). Curve (3) shows the same procedure on another cell but with 5×10^{-6} M cAMP + 2 mM ATP + 3 mM Mg^{2+} added to the perfusing solution.

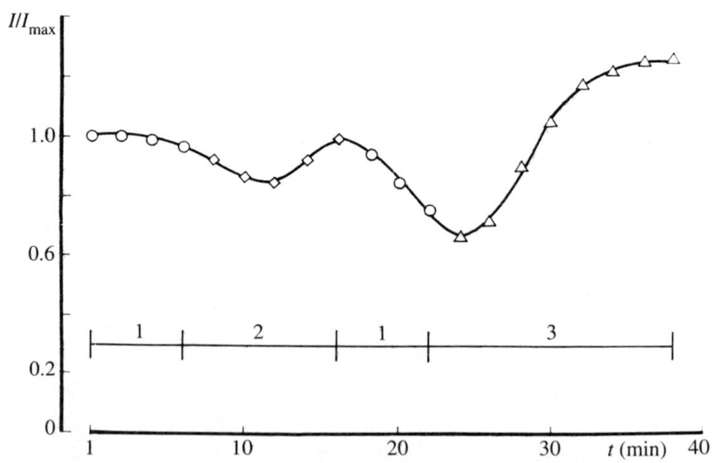

Fig. 6.7. Increase in the restoring effect of cAMP on the calcium current after the introduction of EGTA in the perfusing solution in a snail neuron (Doroshenko *et al.* 1982). The cell was perfused with Tris-aspartate (1), then with a solution containing cAMP + ATP + Mg^{2+} (2), then the same solution with 15 mM EDTA added (3).

considerations has been proved in experiments in which the activity of PhDE has been changed in another way—by specific inhibitors. Intracellular introduction of theophylline always slowed down the decrease of calcium current and its block by intracellular Ca^{2+} (but in this case no recovery of current amplitude could be observed). The participation of calmodulin in these processes can be anticipated from the experiments in which the effects of calmodulin blockers on the 'wash-out' of calcium currents has been studied. Both trifluoperazine and substance R24571 substantially slowed down this process when applied intracellularly (Doroshenko and Martynyuk 1984). Examples of the effects of calmodulin blockers are presented in Fig. 6.8.

In parallel it has been shown that stimulation of proteolysis and a decrease in intracellular ATP level can also influence the calcium current 'wash-out' (Chad and Eckert 1986; Eckert *et al.* 1986; Belles *et al.* 1988). In certain cases the introduction of cAMP into the dialysed neurons did not prevent the 'wash-out', but some positive effect was observed with ATP (Byerly and Yazejian 1986) or AMP (Kononenko and Shcherbatko 1988*a*). A supporting effect of ATP on the functioning of calcium channels has also been observed in smooth muscle cells (Ohya *et al.* 1987). Recently, in the

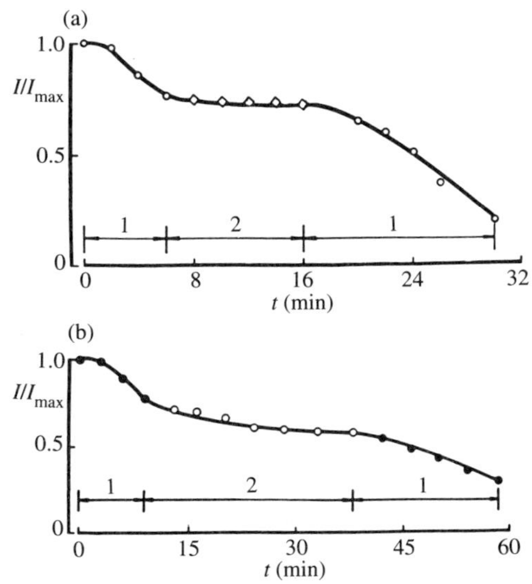

Fig. 6.8. Effects of intracellular administration of calmodulin-blockers on the 'wash-out' of calcium current during intracellular perfusion of a snail neuron (Doroshenko and Martynyuk 1984). The neuron was perfused with basic solution (1) and a solution containing calmodulin-blockers (2): (a) 10^{-5} M trifluoperazine and (b) 10^{-6} M R24571.

cytoplasm of cardiomyocytes a special factor has been found which prevents the 'wash-out' effect in calcium channels; the activity of this factor (having a molecular weight of 20–30 kDa) could be prevented by the application of trypsin or by heating (Kameyama *et al.* 1988). A factor acting in similar way has been isolated from *Paramecia*: injected into mutant species in which calcium channels could not be revealed in the membrane, it 'cured' them and restored the calcium conductance (Haga *et al.* 1984).

It is likely that experimental conditions vary in different laboratories working on different tissues, especially with respect to the calcium buffering capacity of the cells studied; at a higher capacity the effects of the calcium-dependent enzymes (for instance, protein kinases) would be less prominent; this possibility has been discussed in detail by Levitan (1985).

It is also quite possible that in different types of neurons the degree of cAMP-dependent control of calcium channels may be different. This possibility has been stressed in the review of Byerly *et al.* (1988). As already mentioned, in our experiments the positive effect of intracellular cAMP was variable, and in some cells it could not be detected. A definite correlation between its presence and some functional properties of the neuron was found: a substantial potentiation of calcium currents upon intracellular introduction of cAMP was seen in cells in which a similar potentiation could be produced by extracellular application of serotonin (Fig. 6.9). Serotonin-induced potentiation of the calcium current in some snail neurons was also

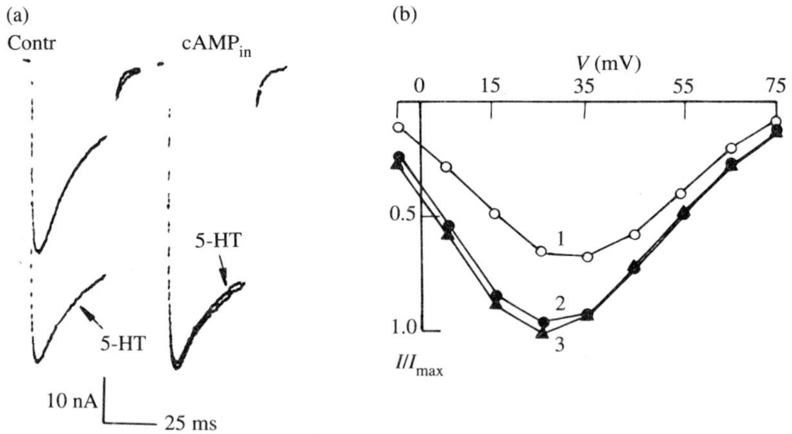

Fig. 6.9. Effect of intracellular administration of cAMP on the potentiating action of serotonin on calcium currents in a snail neuron (Kostyuk *et al.* 1990). (a) Records of calcium currents are shown, obtained during cell perfusion with basic solution (Contr) and a solution containing 10^{-5} M cAMP ($cAMP_{in}$); (b) I–V curves measured in control (1), during external application of 10 μM serotonin (2), and during similar application and intracellular introduction of cAMP (3).

observed by Paupardin-Tritsch *et al.* (1986*a*) and Gerschenfeld *et al.* (1986), and in *Aplysia* neurons by Pellmar and Carpenter (1979) and Lotshaw and Levitan (1988). This potentiation could be reproduced by the action of forskolin, IBMX, and dibutyryl-cAMP (Levitan and Levitan 1988). In our experiments, cells that were insensitive to injection of cAMP also did not respond to serotonin. An example of this, taken from the work of Kostyuk *et al.* (1990), is shown in Fig. 6.10. Possibly, the mechanism of cAMP-dependent phosphorylation in some neurons serves as a mediator of the natural modulatory action of serotonin on calcium channels, and this mechanism is not expressed in cells in which the function of serotonin and the corresponding receptors are absent.

Pronounced changes in the metabolic modulation of calcium channels have been observed in rat sensory neurons during ontogenic development of the animal. The clear supporting effect of intracellular introduction of cAMP shown in cells from new-born animals was much less prominent or even absent in cells from adult animals (Veselovsky *et al.* 1986). It is not yet clear whether we are dealing here with real changes in the modulatory mechanism or with alterations in the accessibility of the inner surface of the membrane to the substances introduced during cell dialysis.

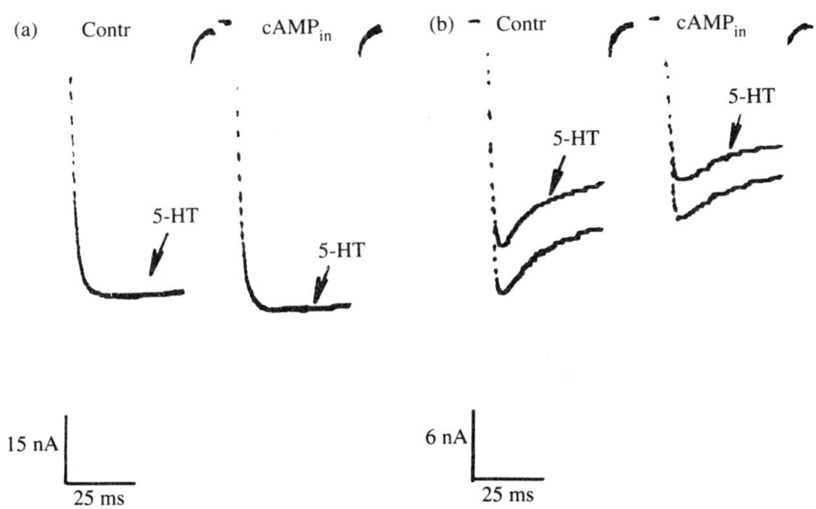

Fig. 6.10. Absence of an effect of intracellular administration of cAMP in snail neurons in which serotonin did not affect, or even inhibited, the calcium current (Kostyuk *et al.* 1990). In (a) and (b) records on the left were obtained during perfusion with basic solution; the ones on the right were obtained when 10^{-5} M cAMP was added to the cell. In (a) the currents before and after application of serotonin overlapped completely.

Adenylate cyclase is a complex system of membrane proteins in which, besides the external receptor and internal hydrolytic components, an important role is played by the intermediate GTP-binding components (G-proteins). Findings about the possibility of both up- and down-regulation of the enzymatic activity, and direct biochemical studies, have revealed that the G-proteins are, in fact, a complex of substances, some of which transmit the activating (G_s) and others the inhibitory (G_i) signals to the catalytic centre from the corresponding receptors. Some G-proteins do not fit in these groups (G_o). Of great importance was the detection of natural toxins which act specifically on certain types of G-proteins. Cholera toxin acts on G_s, stabilizing it by ADP-ribosylation and hindering in this way the subsequent activation of the catalytic centre (Ui *et al.* 1984). Pertussis toxin acts specifically on G_i, inactivating its ability to bind GTP (Jacobs *et al.* 1984).

The best-known example of up-regulation involving G_s protein is the potentiation of high-threshold calcium currents in the cardiomyocytes under the action of β-adrenomimetics. Dibutyryl-cAMP also potentiates calcium currents in astrocytes; isoproterenol is an agonist of this action (MacVicar and Tse 1988). Probably, the same mechanism underlies the increase of calcium currents in snail neurons evoked by serotonin, and in cardiomyocytes evoked by histamine (Hescheler *et al.* 1987*b*). Stimulation of calcium influx by histamine has also been observed in brain synaptosomes; dibutyryl-cAMP and PhDE inhibitors potentiated this effect (Rodriguez *et al.* 1988). The effect of histamine on the somatic membrane has not yet been studied. However, in some structures the cAMP-dependent phosphorylation triggered through β-receptors *inhibited* the calcium conductance (smooth muscle fibres from the mesenteric artery, Saida and van Breemen 1985; tonoplast-free cells from *Nitellopsis*, Shiina *et al.* 1988).

Depression of calcium currents due to changed adenylate cyclase activity through G_i proteins is well known in cardiomyocytes as a consequence of the action of acetylcholine on muscarinic cholinoceptors (Fischmeister and Hartzell 1986; Hescheler *et al.* 1986). There are no direct data about possible down-regulation of calcium conductance in neuronal membranes through the PK-A system, although numerous examples of such regulation under the action of physiologically active substances based on other principles are available (see below).

The cAMP-dependent protein kinase has now been isolated from heart muscle. Its catalytic subunit is a monomer with molecular weight 38–39 kDa, showing a tendency to aggregation at neutral pH and low ionic strength of the medium (Peters *et al.* 1977; Severin *et al.* 1981). Unfortunately, it is not yet clear how phosphorylation of some components of the calcium channel can modulate the activity of the latter. Some time ago, Rinaldi *et al.* (1981) observed cAMP-dependent phosphorylation of a protein with molecular weight 23 kDa that proceeded in parallel with an increase in calcium influx in vesicles obtained from cardiac muscle membranes. The authors named this protein 'calciductin' and suggested that it represents the phosphorylated component of the calcium channels. However, this suggestion met objections. First, it was doubtful whether the influx of calcium into the vesicles really took place through calcium channels. It could also happen through the sodium–calcium exchange system; indeed, its sensitivity to verapamil was about three orders of magnitude less than that of calcium channels (Philipson 1983). Secondly, the preparation could be contaminated by proteins from the sarcoplasmic reticulum, which has its own system of calcium transport. The 'calciductin' actually seemed very similar to phospholamban—the protein which modulates the activity of the calcium pump in the sarcoplasmic reticulum (Hinke *et al.* 1984). After the separation of subunits from the purified calcium channels of the skeletal muscle T-system and their phosphorylation *in vitro*, it has been shown that PK-A phosphorylates both the α_1-subunit (molecular weight 165 kDa) and the β-subunit (molecular weight 55 kDa), respectively (Curtis and Catterall 1984; Hosey *et al.* 1986; Imagawa *et al.* 1987*a*). It has been shown that cAMP-PK phosphorylates the α_1-subunit at the serine residues, and to a lesser extent at the threonine residues. The same subunit can be phosphorylated by Ca-calmodulin-dependent protein kinase (Calm-PK), in this case predominantly at the threonine residues (O'Callahan and Hosey 1988). Recently, the amino-acid sequence of the α_1-subunit from the T-system of skeletal muscle has been determined; the main site of phosphorylation was shown to be serine-687 (Rohrkasten *et al.* 1988).

Further experiments are needed to determine which site in the channel is really phosphorylated *in vivo* and which is important for the changing of the function of the channel; it would also be of importance to compare the data from muscle fibre membrane with those about neuronal high-threshold calcium channels.

Direct modulation of calcium channels by GTP-binding proteins

It has been shown by many investigators that several neurotransmitters depress calcium currents in sensory neurons of the dorsal root ganglia. This effect was first described by Dunlap and Fischbach (1978, 1981) and later by Forscher and Oxford (1984) for noradrenalin. Only high-threshold currents are affected (McFadzean and Docherty 1987), and the secretion of substance P is depressed in parallel (Dunlap and Fischbach 1981). The calcium currents can also be depressed by γ-aminobutryric acid (GABA) (Dunlap and Fischbach 1981; Okamoto *et al.* 1983, Deisz and Lux 1985) and its agonist baclofen (Scott and Dolphin 1986). GTP-binding proteins are involved in these effects, as a similar depression can be produced by intracellular injection of non-hydrolysable GTP analogue (GTP-γ-S, GMP-PNP) which induce a long-lasting activation of the corresponding proteins (Dolphin and Scott 1987, 1989). At the same time, depression is not connected with any changes in the intracellular cAMP level, although it could be blocked by toxins affecting the adenylate cyclase complex (pertussis toxin). A depression of calcium currents was also caused by adenosine, which affects the adenylate cyclase system through A_1-receptors connected in turn with G_i-proteins (Dolphin *et al.* 1986; Macdonald *et al.* 1986).

All these data led to the conclusion that modulation of calcium-channel activity may occur through direct interaction of the membrane G-proteins with the channels. This suggestion has been widely supported and is used now for the explanation of the modulatory action of many neurotransmitters (reviewed by Ewald *et al.* 1988). Nevertheless, the possible existence of transmitter receptors directly on the voltage-operated calcium channels cannot be excluded (Forscher *et al.* 1986).

The depressing effect of noradrenalin (and dopamine) on calcium channels was also shown on neurons from other structures: mammalian brain (Williams and North 1985), snail ganglia (Akopyan *et al.* 1985; Gerschenfeld *et al.* 1986), and sympathetic ganglia (Horn and McAfee 1979, 1980; Marchetti *et al.* 1986). In the latter case the effect is mediated through α-adrenoceptors and can be antagonized by the relevant blockers (phentolamine). A more detailed analysis has shown that this effect is mediated by α_2-receptors (McAfee *et al.* 1981) and could be reproduced in frog sympathetic neurons (Koketsu and Akasu 1982). The effects were not connected with the changes in the intracellular cAMP level, although they were sensitiver to pertussis toxin and to the action of antibodies specific for the α-subunit of the G-protein. Injection of the purified subunit mimicked the inhibitory effect of dopamine (Harris-Warrick *et al.* 1988). A similar mechanism seems to mediate the inhibitory muscarinic cholinoceptive action of acetylcholine on calcium currents in sympathetic neurons (Wanke *et al.* 1987). In this case, the inhibitory action of acetylcholine and noradrenalin could be imitated in a prolonged form by introduction of non-hydrolysable analogues of GTP (GTP-γ-S) into the cell. However, pertussis toxin, which blocks the G_i-protein, removed only the depressive action of acetylcholine without changing the effect of noradrenalin (Song *et al.* 1989); the latter may be achieved by some other intermediates (G_o-proteins ?). The role of G_o-proteins in the down-regulation of calcium currents in nerve cells has been suggested by Rosenthal *et al.* (1989). Depression of the inhibitory effect of acetylcholine by pertussis toxin was also demonstrated in hippocampal neurons (Toselli and Lux 1989). Somatostatin seems to inhibit calcium currents in sympathetic neurons similarly, through a G-protein sensitive to pertussis toxin (Ikeda and Schofield 1989).

The intrinsic mechanism of interaction between G-proteins and calcium channels is quite obscure. It has been shown that GTP-γ-S produce a substantial increase in the activation and deactivation time of calcium channels, and changes the effect of channel antagonists (nifedipine, D-600, diltiazem); all these actions imply conformational changes in channel structure (Dolphin and Scott 1989).

In certain ('bursting') neurons of *Helix pomatia* the calcium currents could also be inhibited by serotonin (Kononenko and Shcherbatko 1985*b*). These experiments were performed on non-perfused neurons in which, despite the application of potassium-channel blockers, some outward current remained; however, the effect could not be a result of their increase, as direct tests did not show any effect of serotonin on potassium currents in these cells. Further experiments have shown that neither is the effect connected with changes in the intracellular cAMP level. Direct changes in the latter in dialysed cells did not affect the inhibitory action of serotonin (Kostyuk *et al.* 1990). Figure 6.10b illustrates this effect of serotonin.

In the same cells the inhibition of high-threshold calcium currents was observed on the background of spontaneous inhibitory post-synapatic potentials generated by the activity of some unidentified interneurons. Part of the change in calcium current amplitude could be due to the decrease in membrane resistance during synaptic action; but after correction for this change the calcium current was still decreased by about half. Similar synapatic responses could be induced in the cell by stimulation of certain peripheral nerves. The transmitter for this action is as yet unknown (Kononenko 1984).

Down-regulation of the neuronal calcium channels has also been produced by endogenous peptides. Met- and Leu-enkephalins, as well as morphine, blocked calcium currents in neuroblastoma × glioma hybrid cell line (Tsunoo *et al.* 1986; Hescheler *et al.* 1987*b*; Shimahara and Icard-Liepkalns 1987; McFadzean 1988). The effect could be abolished by pertussis toxin and restored by intracellular introduction of the G-protein α-subunit; possibly, in this case, the channels were also directly affected by G-proteins. In snail neurons calcium currents could be depressed by cholecystokinin (Hammond *et al.* 1987) and the endogenous FMRF-peptide (Brezina *et al.* 1987). In mice, sensory neuron inhibition could also be produced by dynorphin as well as by enkephalins (Werz and Macdonald 1982, 1984; Macdonald and Werz 1986), which affected predominantly the high-threshold currents (Gross and Macdonald 1987). However, a potentiating effect of morphine on calcium currents in neuroblastoma cells exerted through μ-receptors has also been described (Lorentz *et al.* 1988). Finally, a new endogenous peptide inhibitor of high-threshold calcium currents with a molecular weight below 1 kDa has been isolated from brain tissue; it probably acts by direct binding to the channels (Janis *et al.* 1988). Callewaert *et al.* (1989) also isolated an endogenous peptide of similar size, which exerted a quite peculiar effect: it up-regulated high-threshold calcium channels in cardiac muscle and down-regulated them in neurons.

There are scattered data about possible up-regulation of calcium channels by G-proteins. Scott and Dolphin (1987) observed potentiation of the agonist action of Bay K 8647 on calcium currents after intracellular introduction of a non-hydrolysable GTP-analogue in dorsal root ganglion neurons. Potentiation of calcium currents in hippocampal neurons by noradrenalin was observed by Gray and Johnston (1987); however, it could also be mimicked by injection of cAMP or application of forskolin. Some increase (≅20 per cent) of calcium currents in snail neurons could be induced by application of parathyroid hormone (Kostyuk *et al.* 1990). This effect remained despite progressive 'wash-out' of calcium currents and corresponding decrease in the intracellular cAMP level. Although we do not know if this hormone exists in snail ganglia, there are data about the effective binding of the exogenous hormone there, indicating the possible existence of receptors to a similar polypeptide which may act as a natural agonist of calcium channels.

The up-regulation of calcium channels through G-proteins acting in parallel with indirect modulation via cAMP-dependent phosphorylation has also been demonstrated on cardiomyocytes (Heschler *et al.* 1987*a*, *b*; Shuba *et al.* 1990) and on smooth muscle fibres of the ear artery (Benham and Tsien 1988).

Modulation of calcium channels by other protein kinase systems

After the detection of the Ca-phospholipid-dependent protein kinase (C-kinase) attempts were made to reveal its possible participation in the modulation of calcium channels. DeRiemer *et al.* (1985) found that proteinkinase C (PK-C) activation by phorbol esters potentiates calcium currents in *Aplysia* neurons; Harris *et al.* (1986) also found a parallel increase in norepinephrine secretion in phaeochromocytoma cells. A steady increase of high-threshold calcium current in some snail neurons by phorbol ester has been observed by Doroshenko and Kostyuk (1987); an example of such an action is shown in Fig. 6.11.

Potentiation of calcium influx and insulin secretion in pancreatic B-cells by phorbol esters was described by Bozem *et al.* (1987); however, in natural conditions the modulation of calcium signals in these cells may occur through the cAMP-PK system (Henquin *et al.* 1987). In cultured cardiomyocytes, potentiation of calcium currents during activation of PK-C has also been observed; it could be augmented by membrane depolarization (Lacerda *et al.* 1988). In *Aplysia* bag cells the involvement of the activity of additional calcium channels with higher unitary conductance has been observed after treatment with PK-C activators (Strong *et al.* 1987). Activation of calcium channels has also been observed on clonal osteoblastoma cells (Yamaguchi *et al.* 1987). On the other hand, Hammond *et al.* (1987) have seen an opposite effect after direct injections of PK-C into some snail neurons: an increase of the inhibitory action of cholecystokinin. An inhibitory effect of

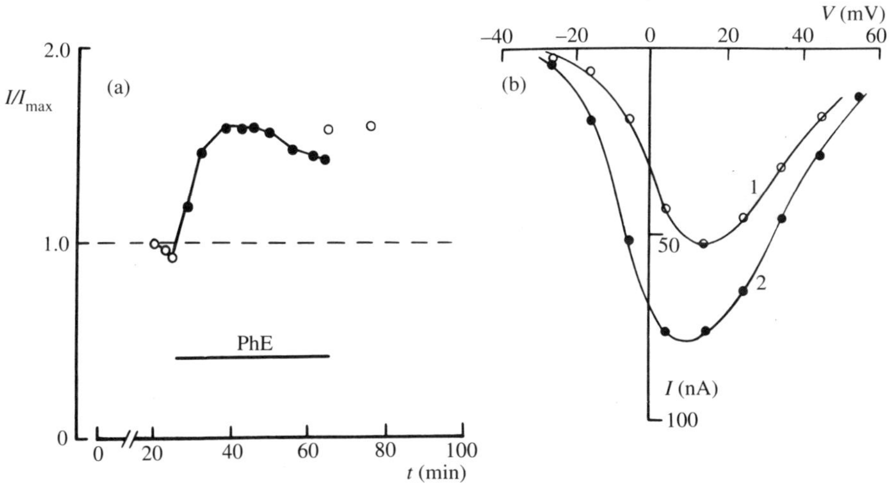

Fig. 6.11. Effect of phorbol ester (PhE) on the calcium current in a perfused snail neuron (Doroshenko and Kostyuk 1987). (a) Normalized calcium current amplitudes during perfusion with basic solution (open symbols) and during external application of 0.05 μg/ml PhE (filled symbols); (b) the corresponding $I-V$ curves before (1) and during (2) application of PhE.

phorbol esters and diacylglycerol on the high-threshold calcium channels in aortic smooth muscle fibres was noticed by Galizzi *et al.* (1987). A depression of both low- and high-threshold components of the calcium currents in clonal pituitary cells and chick sensory neurons by PK-C activators has been described by Marchetti and Brown (1988). Finally, in mice sensory neurons a depression of the inactivating high-threshold calcium current was observed upon addition of either PK-C activators or forskolin and dibutyryl-cAMP (Gross and Macdonald 1988).

Despite plenty of observations about the possible role of PK-C in calcium-channel modulation, the problem of activation of this kinase by some external or internal factors in natural conditions is completely unclear. One may suggest that this mechanism can be switched on in addition to other processes, for instance by elevation of the intracellular calcium level due to the activity of calcium channels or by its release from intracellular stores. In this case, PK-C can act as a supporting mechanism for cellular responses induced by other mechanisms (the so called 'gain control'—see Rasmussen *et al.* 1985).

cGMP-dependent phosphorylation seems to be the rarest mechanism of calcium-channel modulation; it has been described in some snail neurons (Paupardin-Tritsch *et al.* 1986*b*). In isolated hippocampal neurons, a depression of high-threshold barium currents was observed after the intracellular injection of cGMP or external application of dibutyryl-cGMP (Doerner and Alger 1988). However, the authors suggested a possible indirect mechanism of this action: through activation of cAMP-PhDE and decrease of the intracellular level of cAMP.

Figure 6.12 presents in a schematic way the intracellular systems described above which modulate the activity of high-threshold calcium channels.

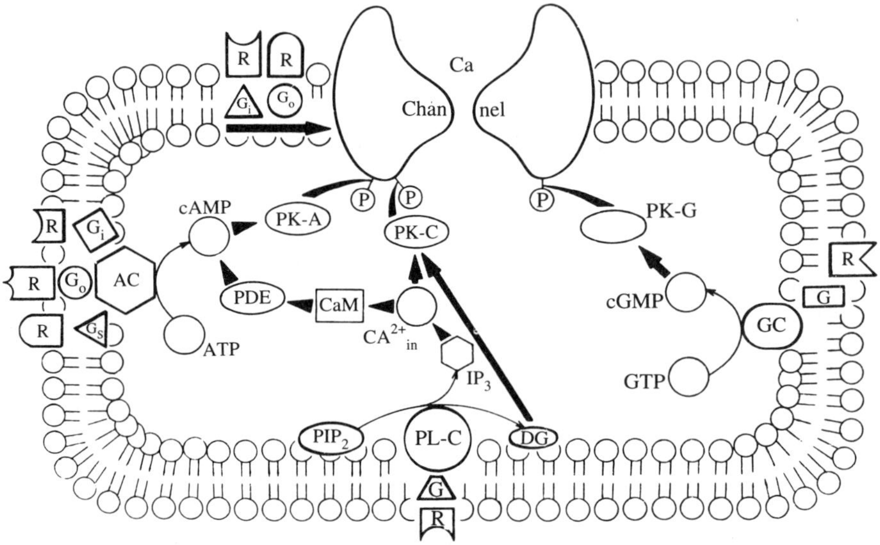

Fig. 6.12. Schematic presentation of the possible cellular mechanisms involved in calcium-channel modulation. R, membrane receptors; G, G_o, G_i, GTP-binding proteins; AC, adenylate cyclase; GC, guanilate cyclase; CaM, calmodulin; PK-A, PK-C, PK-G, different protein kinases; PL-C, phospholipase C. PIP_2, phosphatidylinositol-4,5-bisphosphate; PDE, PIP_2-phosphodiesterase; DG, diacylglycerol; IP_3, inositol-1,4,5-trisphosphate.

Other sorts of calcium-channel modulators

Modulation of calcium channel activity can also be observed under the influence of other substances of quite different structure. Depression of calcium currents can be exerted by alcohols (methanol, ethanol, and butanol; Oyama *et al.* 1986), barbiturates (Werz and Macdonald 1985; Ikemoto *et al.* 1986), and external application of calmodulin antagonists (Bernal *et al.* 1987; Pott and Lipp 1987; Doroshenko *et al.* 1988*a*). In the latter case a positive effect has been observed at low concentrations (below 10^{-6} M). Peculiar effects were observed during application of menthol, amiloride, and furosemide to nerve cells. In chick sensory neurons menthol specifically increased the inactivation of high-threshold calcium channels; in parallel, it blocked the low-threshold ones (Swandulla *et al.* 1987; Doroshenko *et al.* 1989). Furosemide increased the high-threshold calcium current in snail neurons and especially prolonged its 'tail' during membrane repolarization, obviously because of the prolongation of the channel-open state (Kononenko and Shcherbatko 1988*b*). Amiloride seems to be a selective inhibitor of *low-threshold* calcium channels in dorsal root ganglion neurons, which are very resistant to all other blockers compared with the high-threshold ones (Tang *et al.* 1988). Changes in calcium currents of phaeochromocytoma cells were observed when the cells were exposed to lectins (concanavalin A); this effect suggests the existence of a mechanism of signal transmission from the carbohydrate membrane groups interacting with lectins and the calcium channels (Greenberg *et al.* 1987). Finally, in sympathetic neurons an inhibitory effect of prostaglandin E_1 on the calcium component of the action potential was described (Mo *et al.* 1985).

Conclusion

Although the influx of calcium ions into nerve cells takes place mainly through the voltage-operated calcium channels, it can be widely modulated by different neurotransmitters and hormones. The use, for this purpose, of the mechanism of cAMP-dependent phosphorylation of the channel-forming protein is less prominent here compared with cardiac and skeletal muscle fibres. In some neurons the activation of this mechanism induces a considerable potentiation of high-threshold calcium currents; serotonin and some hormones may be the natural agonists for this potentiation. In other neurons such potentiation cannot be revealed; its role may also change during cell ontogenesis. Modulation of calcium channels may also occur more directly via interaction with GTP-binding proteins from the adenylate cyclase complex. Down-regulation of channel function under the action of different neurotransmitters (noradrenalin, serotonin, GABA), endogenous peptides, and other physiologically active substances coupled to the activation of G_i or G_o proteins is most common; up-regulation coupled to the activation of G_s is much more seldom. Calcium channels can also be modulated in some cases through the activation of protein kinase C, and very seldom via cGMP-dependent kinase; these mechanisms may have an auxiliary role in the maintenance of responses induced by other mechanisms (gain control).

7. Development of calcium conductance during ontogenesis

History

It has been established that the presence of specialized ion-conducting structures, ion channels, in the surface membrane is one of the primordial properties of living cells; they can, for example, be clearly demonstrated in the oocytes of invertebrates. Microelectrode studies of such convenient objects as oocytes of Tunicates and Echinodermata showed the presence of membrane potentials of small value (several millivolts). At such a potential a cell was not capable of generating active responses, but if the membrane was hyperpolarized by electric current, action potentials appeared upon removal of the hyperpolarizing current, which indicates the presence of electrically operated ion channels in the cell membrane (Miyazaki *et al.* 1974). The study of the dependence between action potentials and ionic composition of the medium, as well as the action of different toxins, showed that, just as in differentiated cells, the action potential of oocytes has 'sodium' and 'calcium' components, i.e. several types of ion channels are involved in its generation (Miyazaki *et al.* 1975).

The presence of electrical excitability was then demonstrated in unfertilized egg cells of other animals; the oocytes proved to be very convenient material for the use of the voltage-clamp technique. This soon led to extensive studies of different transmembrane (including calcium) currents. The changes observed in calcium currents during fertilization and subsequent differentiation were of great interest. The study of the former was methodically comparatively simple. However, it was more difficult to obtain data during differentiation because it is not easy to find a species in which the changes in membrane mechanisms could be followed from the stage of oocyte to a cell of differentiated tissue. Some artificial approaches helped to solve this problem. One of these is the use of substances that prevent cell division but do not affect differentiation. In the oocytes of some invertebrates (Polychaeta, Ascidia) such differentiation can be induced parthenogenetically by placing them into a solution with increased K^+ content. As a result, the oocyte develops a differentiated but unicellular embryo, which permits a detailed study of the properties of its membrane during development (Hagiwara and Miyazaki 1977).

To investigate the different stages of cell differentiation, reproduction in cell culture has been used when, depending on time of cell development, one can observe particular changes in cell structure and study peculiarities of ion membrane conductance. Spitzer and Lamborghini (1976) were the first to explant dissociated cells from the neural tube of a tadpole and observe their morphological development and changes in action potential mechanisms. Since then, cell cultures of other animals, including mammals, have been used successfully.

Finally, clonal pathologically transformed cells have also been used for this purpose, as their morphological characteristics can readily be changed by modification of the culture conditions. Especially popular in this respect are cells from the neuroblastoma tumour which were cloned in 1969 (Nelson *et al.* 1969) and have since been widely used in experiments. With this model it is possible to carry out parallel measurements of structural changes and the changes in ionic conductance of the cell membrane.

Calcium currents in the membrane of oocytes

The first analysis of calcium currents in the oocyte membrane was carried out on starfish by Hagiwara *et al.* (1975) and on tunicate by Okamoto *et al.* (1976*a*, *b*). Besides the low-threshold sodium inward current they detected an inward current component in the membrane which depended on the presence of Ca^{2+} ions in the extracellular solution. The I–V curve for this component was quite different from that for the sodium current; it could also be carried by Ba^{2+} and Sr^{2+} ions and it saturated at high concentrations of carrier ions in the intracellular medium. Blockers of calcium currents in differentiated cells (Co^{2+}, Mn^{2+}) also inhibited this component in oocytes.

The use of intracellular dialysis allowed Takahashi and Yoshii (1978) to separate the tunicate oocyte calcium current in a pure form and to analyse it in detail. In all characteristics it appeared to be identical to the high-threshold calcium currents in differentiated excitable cells, indicating that the corresponding channel-forming proteins are, in both cases, expressed from the same basic genetic programme.

Later, similar (or slightly different) results were obtained on oocytes from other animals, including mammals. The presence of calcium inward current channels has been shown in oocytes of Polychaeta (Hagiwara and Miyazaki 1977). In mouse oocytes action potentials have been recorded that are based on the activation of calcium channels similar to those in differentiated cells in their potential-dependence and pharmacological sensitivity (Yoshida 1982). In calcium-free solutions they were converted into sodium-conducting channels (see Chapter 2); but retained their sensitivity to calcium-channel blockers.

The functional meaning of the presence of calcium channels in oocytes may be related to the process of fertilization and the subsequent induction of morphological differentiation. During fertilization a characteristic membrane potential change occurs (the so-called activation potential). In marine worm oocytes the membrane potential changes at this stage from −33 to +51 mV. This prolonged activation potential can last for several minutes and is due to a stationary increase in the sodium conductance of the membrane. Membrane depolarization activates pre-existing calcium channels and induces calcium influx in the cell; the elevation of $[Ca^{2+}]_{in}$ may be the signal that triggers further developmental changes (Jaffe *et al.* 1979). Similar changes were observed in ascidian oocytes (Kozuka and Takahashi 1982). However, during fertilization of mouse oocytes only an increase of calcium currents was observed without the steady elevation of sodium conductance (Yamashita 1982).

Whereas calcium channels are 'primordial' ion-conducting structures in the oocyte membrane, the presence of sodium channels here is much more variable. In starfish and tunicate oocytes in which a sodium inward current component is always recorded even before activation, it nevertheless differs substantially from sodium currents in differentiated cells. It depends to a certain extent also on the presence of divalent cations in the external solution and can be blocked by Co^{2+} ions, like the calcium current. This may indicate that sodium channels in these oocytes have a lower selectivity and may also pass calcium into the cell.

Calcium currents during natural differentiation of nerve cells

In oocytes that show only high-threshold calcium currents before activation (Polychaeta), the use of the 'differentiation without cleavage' technique has clearly demonstrated that the appearance of ion channels of other types is a secondary process. In the first 4–5 hours after activation such cells acquire the ability to make amoeboid movements; at this time potassium channels appear in their membrane. Sodium channels show up later when the embryo acquires cilia and become very vivid (Hagiwara and Miyazaki 1977).

An artificial arrest of cleavage can also be induced during the later stages of development (by the use of cytochalasin). Takahashi and Yoshii (1981) induced such arrest at the 16-cell stage of the development of an ascidian embryo. In this case the sodium and calcium currents first decreased (for 10 hours) and then started to increase; their *I–V* curves became shifted towards more positive membrane potentials (probably because of a change in fixed charge density on the membrane surface). If cleavage was arrested at the 64-cell stage, the cells that will further produce the ectodermal and muscular layers could already be identified. In the former cell type both sodium and calcium currents could be recorded, and in the latter only calcium currents. All these changes in the interrelations of different currents reflected the changes only in the density of corresponding channels but not in their properties (for instance, selectivity). Thus, the synthetic apparatus of the cell can already produce proteins which form sodium and calcium channels in differentiated cells at the earliest stages of ontogenic development, and only the expression of this possibility varies during the course of cell differentiation.

In the laboratory of Lazdunski, specific blockers of sodium (TTX) and calcium (nitrendipine) channels have been used for the investigation of the ontogenesis of the corresponding channels (Kazazoglou *et al.* 1983; Lombet *et al.* 1983). In these experiments it has been shown that the density of TTX receptors in rat brain and skeletal muscle increases immediately before the birth and reaches a stationary level after 10–12 days of postnatal development. A similar dependence was found for the cerebellum; however, in this case the density of TTX receptors started to increase 10 days after the birth and reached a steady level after 30 days. It should be mentioned that the dissociation constants for TTX in muscle did not differ during the pre- and postnatal periods, whereas in brain and cerebellum they gradually increased and by 10 days after the birth became respectively 10 and 3 times larger than those during the prenatal period. The density of nitrendipine receptors changed in a similar way, but in this case the K_d values did not change. The density of these receptors was identical in heart, skeletal muscle, brain, and cerebellum 5 days before birth and then started to increase. The rate of rise increased in the order cerebellum < brain = muscle < heart, and their steady level in the postnatal period was lowest in the cerebellum. In the cerebellum the nitrendipine receptors developed in parallel with the TTX and GABA receptors, and in heart the β-adrenoceptors appeared before the nitrendipine ones.

The investigation of ionic currents in neuroblasts developing in culture has shown similar changes. In dissociated cells obtained from the tadpole neural lamina, only calcium action potentials could initially be recorded; later, a sodium component also appeared. At even later stages many cells showed the action potential of a purely 'sodium' nature (Spitzer and Lamborghini 1976).

More extensive data about the ontogenesis of calcium channels have been obtained on a very convenient tissue—sensory neurons from dorsal root ganglia isolated before birth, immediately after birth, and at different periods of postnatal development. Initially, Matsuda *et al.* (1976, 1978) found that many cells taken from new-born animals can generate action potentials in sodium-free solutions. With ageing, the proportion of such cells decreased (Yoshida *et al.* 1978; Yoshida and Matsuda 1980). Therefore the authors concluded that calcium conductance is more characteristic for undifferentiated cells. As they did not measure transmembrane currents, they could not determine the quantitative and qualitative characteristics of this conductance, especially the type of calcium channels involved. Therefore in our laboratory more detailed studies of ionic currents in rat sensory neurons during their ontogenesis have been performed, using the intracellular perfusion technique for reliable current separation (Fedulova *et al.* 1986; Kostyuk *et al.* 1986).

For the investigation of the changes appearing during postnatal development, neurons were isolated from animals of three age-groups: 8–10, 45–50, and 90–100 days after birth; in each group 76 cells were studied. The densities of different inward current components were determined; they were evaluated as the ratio of maximal current amplitude to the capacitance of the membrane (the latter was measured by automatic integration of the area under the curve of the capacity charge current). The total inward current triggered by membrane depolarization could be separated into two components, differently expressed in different cells: the 'fast' TTX-sensitive and 'slow' TTX-resistant sodium, low- and high-threshold calcium ones. A comparison of the activation and inactivation time-constants for each component in neurons of each age-group has shown that the dispersion of parameters within each group was comparable to their dispersion during transition from one group to another. Therefore, it is likely that during postnatal development there are no significant changes in the potential-dependence of the stationary and kinetic characteristics of all the four components of the inward current; this means that the properties of the corresponding channels also do not change.

When investigating quantitative changes in currents, it is important to take into account the variations in the expression of their components in different cells. We used as a criterion of the presence of a certain type of channel the generation of the corresponding current with an amplitude above 100 pA; in this case all cells could be separated into six subpopulations according to the expression of different types of channels. In three of these subpopulations a statistically significant change in the number of cells was detected during development. The presence of cells demonstrating all four kinds of channels definitely decreased with age, whereas the presence of cells with only TTX-sensitive and high-threshold calcium channels increased. A statistical analysis with the help of Fisher's ϕ-transformation has confirmed the significance of these changes. A comparison of these data with the distribution of studied cells, according to the diameter of their somata, indicated that cells in which only the TTX-sensitive sodium and high-threshold calcium channels remained were not the largest in size; however, they also did not include the smallest ones. A comparison of the mean density of each current component in different age-groups showed a definite decrease in the density of calcium currents in this subpopulation.

The detection of definite correlations between the densities of different current components was noteworthy. Thus, in the subpopulation showing all four current types, a direct correlation between the densities of TTX-sensitive sodium and high-threshold calcium currents was found; in the subpopulation with only TTX-sensitive sodium and high-threshold calcium currents (the presence of which increases with age) a reciprocal relation between their densities could be demonstrated. Examples of such correlations are presented in Fig. 7.1.

The findings indicate that during the postnatal development of sensory neurons (at least in the rat) quite complicated changes in the quantitative and qualitative expression of different voltage-operated ion channels occur. Already at birth most neurons have TTX-sensitive sodium and high-threshold calcium channels in their membrane; during postnatal development they retain their dominating position. However, another process takes place in parallel: a decrease in the proportion of neurons that show special types of channels—the TTX-resistant sodium and low-threshold calcium ones. At the age of approximately three-months, when the functional differentiation of sensory neurons is obviously accomplished, about 60 per

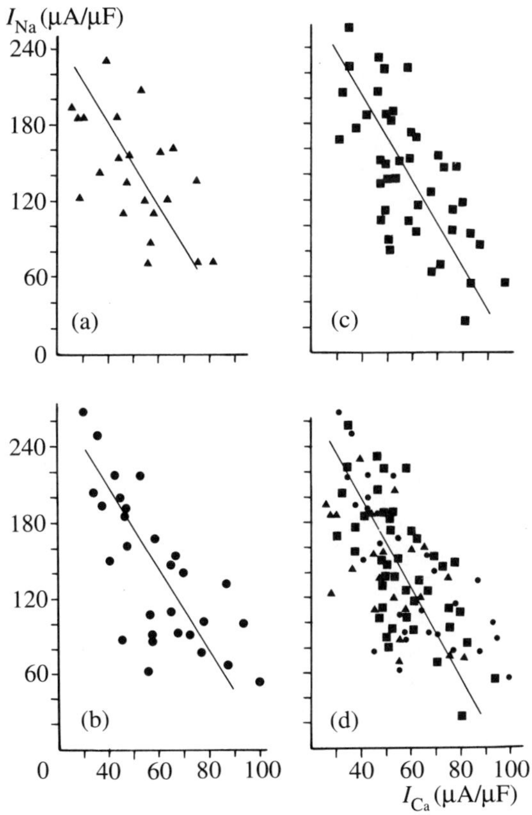

Fig. 7.1. Correlation fields for the densities of I^{f}_{Na} and I^{h}_{Ca} in dorsal root ganglion neurons measured 10 (a), 45 (b), and 90 (c) days after birth; in (d) all age-groups are combined (Fedulova *et al.* 1986).

cent of cells retain only TTX-sensitive and high-threshold calcium channels in their membrane. Thus, the high-threshold calcium channels represent an obligatory component of electrically excitable cell membrane at all stages of their ontogenic development, although their density slowly decreases with age. At the same time, the TTX-resistant sodium and high-threshold calcium channels are mostly the property of young developing neurons (at least in the dorsal root ganglion) and may be related to their functional properties in this period (see Chapter 9).

It is worthwhile to continue this analysis backwards and to discover the changes in channel expression during prenatal development. This is more difficult because of the small size of the neurons and their high vulnerability in the course of an experiment. However, using cell dialysis with a glass micropipette, Fedulova and Veselovsky, in our laboratory, have managed to compare the inward current components in the same neurons taken from 7-day-old embryos and from animals during the first postnatal day. A very peculiar dependence was found. The density of TTX-sensitive sodium and high-threshold calcium currents steadily increased in the prenatal stage, the increase in the density of the calcium current being somewhat less rapid. However, the density of the low-threshold calcium current was quite low 7 days before birth and showed a large maximum immediately after birth, obviously suggesting an important role of these channels at the earliest stages of postnatal development.

All these findings are summarized in Figs 7.2 and 7.3, which illustrate the changes during pre- and postnatal development in the proportion of neurons with different combinations of inward current channels and in the densities of the corresponding currents.

Comparable data have been obtained by Yaari *et al.* (1987) during an investigation of the development of calcium conductance in chick neurons. Embryonic cells from chick hippocampus developing in culture initially expressed only low-threshold metabolically independent calcium channels, while at later stages the high-threshold metabolically dependent channels also appeared (in parallel with the development of cellular processes). In cultured cortical neurons taken from very young animals, both low- and high-threshold (DHP-sensitive) calcium channels could readily be detected; in cultured cells from adult animals only high-threshold channels were present (Dichter and Zona 1989). In cultured embryonic hippocampal neurons, the low-threshold calcium current appeared about 6 days after culturing, when the cells already expressed intensive high-threshold currents (Meyers and Barker 1989).

This dependence on age is clearly expressed during embryogenesis of skeletal muscle fibres. In neonatal fibres both types of calcium channels were present, but during subsequent development the density of low-threshold channels decreased, and in about 17 days they disappeared completely. Denervation did not prevent their disappearance but arrested the increase in

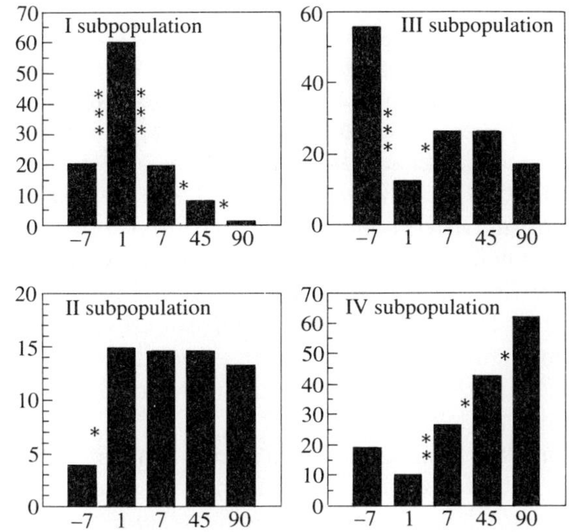

Fig. 7.2. Changes in the proportion of rat dorsal root ganglion neurons, revealing different combinations of inward current channels during ontogenesis. The subpopulations revealed $I^{f}_{Na} + I^{s}_{Na} + I^{l}_{Ca} + I^{h}_{Ca}$ (I), $I^{f}_{Na} + I^{s}_{Na} + I^{s}_{Ca}$ (II), $I^{f}_{Na} + I^{l}_{Ca} + I^{h}_{Ca}$ (III), and $I^{f}_{Na} + I^{s}_{Ca}$ (IV). Abscissa, days before and after birth. Asterisks indicate 10 (*), 5 (**), and 1 (***) % significance levels.

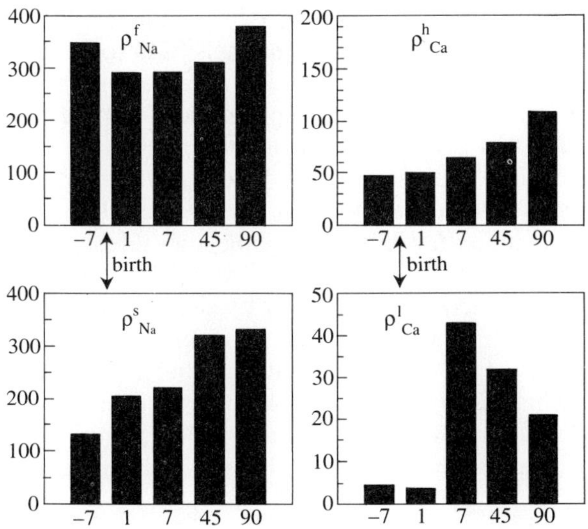

Fig. 7.3. Changes in the mean densities of different inward currents in dorsal root ganglion neurons during ontogenesis. Abscissa, days before and after birth; ordinate, densities of currents in $\mu A/\mu F$.

density of high-threshold channels; muscle isolation performed on the eighth day stopped the disappearance of the low-threshold channels. Thus, it seems that some extrinsic influences regulate these processes (Gonoi and Hagesawa 1988). Approximately the same changes were observed during cultivation of embryonic muscle fibres, which initially have both types of calcium channels. Later, the low-threshold channels remained for a certain period (up to 3 weeks) in the membrane but were not related to the excitation–contraction coupling mechanism. On the contrary, the high-threshold channels increased in number and migrated to the tubular system where they became a component of the mechanism triggering contraction (Beam and Knudson 1988; Romey *et al.* 1989). Finally, even in fibroblasts, which also show two types of calcium channels, the expression of low-threshold but not that of high-threshold channels was depressed during oncogene-induced transformation (Chen *et al.* 1988).

It is noteworthy that a similar situation occurs in a quite different organism, Infusoria, during their life cycle after division. The young organism with low motor activity already has two types of calcium channels in its membrane, but of low density. During maturation and with an increase in motor activity, the density of high-threshold calcium channels also increased, whereas that of low-threshold ones did not change (Deitmer *et al.* 1986).

On the other hand, in some cases the low-threshold calcium channels remain well expressed in completely differentiated neurons or even become predominant. This was shown by Akaike *et al.* (1989) for many neurons from the ventromedial hypothalamus of adult rats and by Crunelli and Pollard (1988) for lateral geniculate neurons of cat and rat. Probably, in these structures the low-threshold calcium channels fulfil some functions specific for mature, differentiated neurons (see also Chapter 9).

It should be mentioned that the expression of low- and high-threshold calcium channels can also be related to such factors as the conditions of culturing. According to the data of Bossu *et al.* (1989*a*), Purkinje cells cultivated in Dulbecco's medium showed both types of calcium currents, whereas during cultivation in Leibowitz's medium the high-threshold channels disappeared.

There are no detailed characteristics of changes occurring in neuronal calcium conductance during ageing. Probably, a decrease of calcium conductance takes place at this time; this has been shown, for instance, by findings about changes in calcium currents in cerebellar granular cells of 24–26-month-old rats (Reynolds and Carlen 1989). A possible reason for such a decrease may be a steady elevation of the intracellular free calcium level in aged neurons.

Calcium currents during artificially induced differentiation

Another technical approach to the problem of ontogenesis of the neuronal calcium conductance is based on the use of dissociated cell cultures and clonal cell lines in which morphological differentiation (growth of neurites, formation of growth cones, establishment of interneuronal contacts) can be induced in some way. As such changes can develop under such external influences very rapidly (a few days), the related changes in the density and function of ion channels can also be determined more precisely.

Some information on this question was obtained some time ago during investigation of the ionic dependence of action potentials in mice dorsal root ganglion neurons cultivated in artificial media (Fukuda and Kameyama 1979). The proportion of cells that reveal a 'calcium' action potential remained constant up to the seventeenth day in culture; however, the rate of rise in the spike increased in parallel with the increase in the growth of neurites. On day 5–6, when the growth of neurites was maximal, the rate of rise was also the largest; it decreased to the initial level on the thirtieth day in culture. The kinetics of 'sodium' spikes did not change during this development. Therefore, the authors came to the conclusion that the growth of neurites is related to an increase in the density of calcium channels in the neuronal membrane (or with an increase in channel conductance, which is less probable).

A similar conclusion was also reached during microelectrode investigations of Purkinje cells in surviving brain slices. The action potentials in the soma of these cells also consist of a 'sodium' and 'calcium' components, whereas in the dendrites they are purely 'calcium' in nature and have a quite prolonged time-course (Llinas and Sugimori 1980*a*, *b*). If cells were investigated in slices taken from animals at different stages of postnatal development, a maximal value of calcium conductance was observed during the period of maximal growth of dendrites (Traub and Llinas 1979). On the basis of these findings, Llinas (1979) has formulated an hypothesis about an 'isometric calcium-dependent growth', according to which intracellular calcium ions are necessary for the incorporation of membrane fragments synthesized inside the cell into the surface membrane. As, according to the findings, the calcium conductance is most strongly expressed during neurite growth, it should also be quite prominent in the membrane of growth cones, supporting here the incorporation in the newly formed membrane of the membrane fragments transported by the mechanism of axoplasmic transport. When the growth of the neurites stops, the growth cones are in fact transformed into synaptic terminals in which the growth becomes isometric (this means that the membranes of synaptic vesicles are now inserted into the surface membranes only to be immediately expelled from it during the turnover of such vesicles). Nevertheless, this isometric growth is still governed by Ca^{2+} ions entering the terminal through calcium channels. A technically elegant confirmation of the possible increase of calcium-channel density in growth cones was obtained with the help of voltage-dependent membrane probes (Grinvald and Farber 1981). The authors used for the excitation of the dye a highly focused beam of a helium–neon laser; the diameter of the spot could be varied between 1.5 and 120 μm. The experiments were made on differentiating neuroblastoma cells that actively produce neurites, and the laser beam was focused on the terminal parts of the latter; the cells were placed in a sodium-free tetra-ethylammonium (TEA)-containing solution. It has been suggested that under such conditions during direct electrical stimulation of neurites the changes in membrane potential that appear (recorded by the probe) would reflect just the inward calcium currents. In several cells the authors managed to show the appearance of local calcium-dependent responses in the region of growth cones or nearby. Along the corresponding neurites such responses could be absent but they appeared again in the cell soma. On the basis of these findings it has been suggested that in a growing neurite the expression of new calcium channels takes place predominantly in the membrane of the growth cone; the calcium conductance in this structure may play a specific role in the growth process regulating membrane neogenesis and the incorporation of protein macromolecules supplied from the soma via the intracellular transport system. However, it should be kept in mind that this technique gives much less precise results than direct measurement of calcium currents; in short neurites it may be difficult to discriminate between a local active response at a chosen point and electrotonic membrane potential changes spreading from the place of stimulation. Therefore such results needed further confirmation.

Direct recordings of calcium currents have been made on large growth cones of *Aplysia* neurons in culture by Belardetti *et al.* (1986). No significant I_{Ca} was detected. However, a calcium current was found in these growth cones by Marom and Dagan (1987), but its density was quite small (corresponding to about one channel per μm^2). In isolated growth cones from PC-12 cells differentiating under the action of nerve growth factor (NGF), the density of I_{Ca} was found to be the same as in cell soma (Streit and Lux 1987). Somewhat contrasting data were obtained on growth cones from sympathetic neurons during measurements of $[Ca^{2+}]_{in}$ with Fura-2, which was found to be significantly increased above the resting level (Lipscomb *et al.* 1988*b*).

In some other experiments the hypothesis about the connection of calcium currents with the neurite growth process did not find support. Scott and Edwards (1980) have found, during culturing of mice dorsal root ganglion neurons, that the proportion of cells that generate 'calcium' spikes decreases, rather than increases, upon prolongation of the time in culture. The authors suggested that, in this case, a possible dedifferentiation of neurons takes place because of the absence of trophic influences from surrounding structures. Dichter *et al.* (1983) observed, in a dissociated culture of rat cortical neurons, a larger calcium membrane conductance component in young cells (3–4 weeks in culture); during maturation it became smaller.

In our group, detailed investigations of ionic currents were performed on non-differentiated and differentiated neuroblastoma cells. Usually these clonal cells have a spherical shape and proliferate continuously. However, under the influence of certain factors (decrease in the amount of serum in the culture medium, increase in pH, etc.) the mitoses stop, and the cells start to develop branching processes similar in shape to neuronal dendrites. This change is reversible to a certain stage, and after removal of the above factors the cells start to dedifferentiate; but after more prolonged influence the restoration of proliferation becomes impossible.

A very convenient method for the regulation of differentiation of neuroblastoma cells turned out to be the pH change of the culture medium. An increase in pH to 7.8–8.2 divided the cell population into two subpopulations: cells that continued to proliferate and retained their spherical shape, and cells that increased in size and started to form processes. After 7–8 days in culture they reached a mature stage, which was characterized by maximal size (up to 60–80 μm in diameter) and an expanded system of the neurite-like processes, reaching 500–1000μm in length. In this state the cells could survive for a long time (several weeks). For the selection of cells for electrophysiological measurements a simple and highly reliable criterion can be used, their size (Pogorelaya and Fomina 1984). Both non-differentiated and differentiated cells were studied using intracellular perfusion for reliable separation of sodium and calcium inward currents; their density was calculated as the ratio between maximal current amplitude and electrical capacity of the cell membrane. It was found that the density of calcium currents in differentiated cells is not significantly different from that of non-differentiated ones. However, the density of sodium currents in differentiated cells was definitely higher; this increase began at the earliest stages of development when the cells had only just started to form neurites. Later, the density of sodium channels did not change much (Veselovsky and Fomina 1986). Examples of such changes averaged from 15 cells are shown in Fig. 7.4.

Peculiar changes of sodium and calcium current densities were observed when neuroblastoma cells that had just started to differentiate were subjected to alternating influences that induce differentiation or stimulate the return to proliferation. During stimulation of proliferation (by a decrease in pH of the medium) a statistically significant decrease in the density of sodium currents, and increase in the density of calcium currents, was observed; during stimulation of differentiation opposite changes occurred (see examples in Fig. 7.5).

Changes in the expression of different types of calcium channels also occur during differentiation of neuroblastoma cells. As already mentioned,

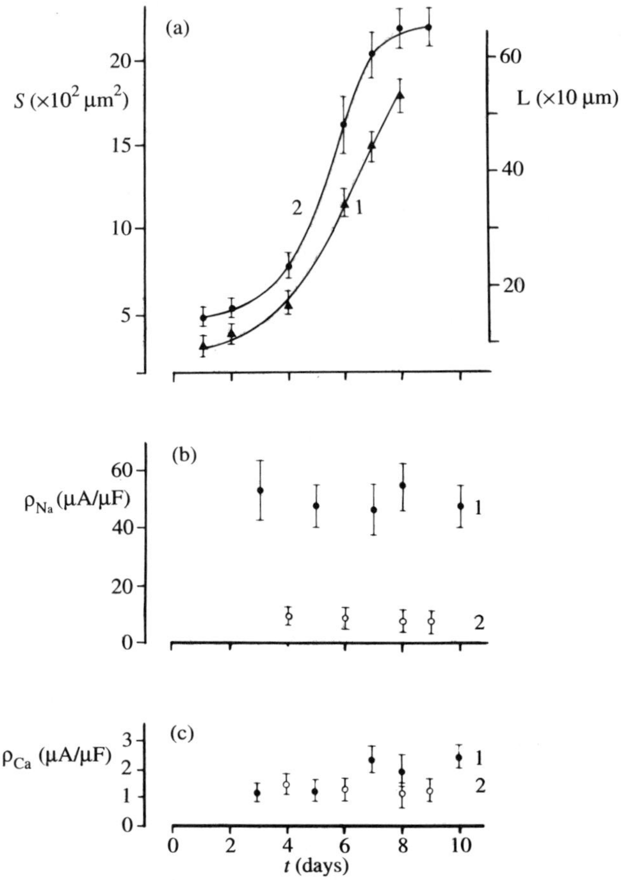

Fig. 7.4. Changes in the mean densities of sodium and calcium currents during differentiation of neuroblastoma cells (Veselovsky and Fomina 1986). (a) Changes in the length of neurites (1) and cell size (2) in the course of differentiation; densities of (b) I_{Na} and (c) I_{Ca} in differentiating and non-differentiating cells.

in non-differentiated cells (which may be considered as analogues of neuroblasts) the calcium conductance is mainly due to low-threshold calcium channels (Kostyuk *et al.* 1988*b*). Because of the low level of membrane potential, these channels are mostly in an inactivated state, and to reveal their presence a hyperpolarization of the cell membrane is needed. During induced differentiation, when the cells start to generate neurite-like processes, an increase in the density of high-threshold calcium currents occurs and the cells acquire the ability to generate 'calcium' spikes (Bodewei

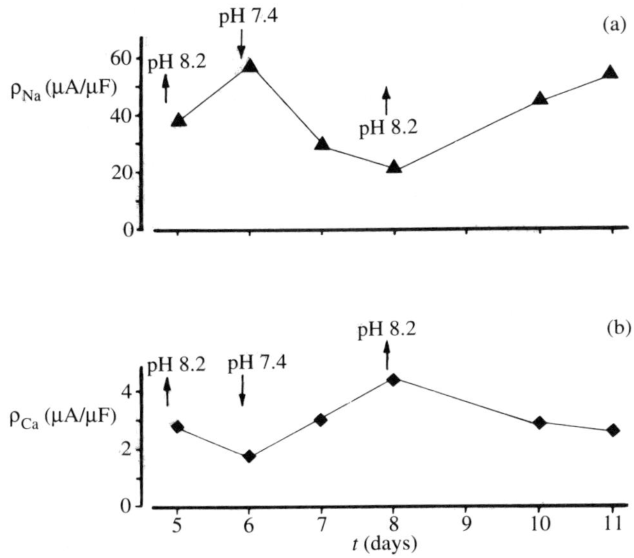

Fig. 7.5. Changes in the densities of sodium (a) and calcium (b) currents in neuroblastoma cells during changes in cultivation medium which promote or depress differentiation (Veselovsky and Fomina 1986).

et al. 1985; Narahashi *et al.* 1987). In other words, during artificially induced differentiation the neuroblastoma cells express changes in the ratio between different types of calcium channels similar to those occurring during natural ontogenic development of neurons.

Just as with the changes occurring during differentiation of healthy neurons, no changes in the characteristic properties of ion-channel functioning have been detected in the course of induced differentiation of neuroblastoma cells; only the number of functional channels in the cell membrane was changed.

Conclusion

During ontogenic development of nerve cells the voltage-operated calcium conductance of the surface membrane seems to be their primordial property. It is already present in unfertilized oocytes. The activation of calcium conductance plays an important role in the complex chain of morphogenetic processes triggered by fertilization. Calcium channels in the oocyte membrane do not differ in their properties from those in the membranes of differentiated neurons; the genetic programme for the synthesis of the corresponding proteins thus remains unchanged during the whole period of the development of the multicellular organism. At different stages of this development only the expression of this programme varies, as shown by the changing density of different types of calcium channels in the cell membrane. Qualitative changes in the mechanism of calcium conductance also occur, mainly in the form of replacement of low-threshold calcium channels, which are more characteristic for developing neurons, by metabolically dependent high-threshold channels, which constitute the main component of calcium conductance in mature cells. However, in some brain neurons the low-threshold calcium channels still retain a prominent functional role once cell differentiation is complete.

8. Fate of calcium ions in the nerve cell

History

The existence in excitable cells of effective mechanisms for controlling the entry of calcium ions into the cell must be connected with the presence of opposite mechanisms which can remove them after the end of cell activity. If the injected ions remained in the cytoplasm in a free state, the functions of the cell would rapidly deteriorate because calcium ions interact with most important macromolecules and change their properties. Thus, an effective system for the maintenance of calcium homeostasis must also exist, probably even more important than the systems maintaining the homeostasis of sodium or potassium ions.

Analytical determination of the calcium content of the cytoplasm was first achieved long ago; initially this was done on the axoplasm extruded from squid axons. The results depended largely on the condition in which the axons were kept. If they were stored in the cold, the axoplasm contained more calcium (about 400 μM; Keynes and Lewis 1956) than in the case of freshly dissected axons (about 50 μM; DiPolo *et al.* 1976). Obviously, this difference represents a continuous calcium enrichment of axons kept in abnormal conditions.

However, these values must mainly represent the calcium bound by some intracellular buffer systems; the level of free (ionized) calcium must be much lower. The first approximations of the ionized part of intracellular calcium were also made on giant axons by Hodgkin and Keynes (1957) from measurements of the movement of injected radioactive calcium in an electric field; the value did not exceed 10 per cent. More precise measurements were hindered by the absence of techniques for the determination of the intracellular activity of ions without destruction of cellular structures. New possibilities were opened by the invention of specific calcium indicators which could be injected into an intact neuron. The first such indicator was the calcium-binding protein, aequorin, which emits light after binding (Shimomura *et al.* 1962). Although the results obtained with aequorin are not free from some errors (dependence of the response on the ionic strength of the medium; the presence of other divalent cations, especially Mg^{2+}; its non-linearity with increase in Ca^{2+} concentration), use of aequorin allowed one to measure Ca^{2+} levels down to 10^{-8} M.

Later, substances were used that absorb light after Ca^{2+} binding; the best known is arsenazo III (a derivative of 4,5-dihydroxy-naphthalen-

2,7-disulphonic acid). This reaction can also induce errors. Because of the high effectiveness of binding and low Ca^{2+} content in the cell, the injection of arsenazo III can bind up to 80 per cent of these ions; therefore, the changes in light absorption may not be directly related to changes in intracellular ionized calcium. Depending on the relation between arsenazo III and Ca^{2+}, complexes of different stoichiometry can be formed; this reaction depends on the pH of the medium and the presence of other divalent cations (Ohnishi 1978). Nevertheless, the use of arsenazo III allowed numerous measurements to be made of the changes in calcium activity in different cells.

Finally, the most effective fluorescent reagents constructed on the basis of calcium chelators (EDTA) were proposed (Quin-2, Fura-2, etc.). They can penetrate through the cell membrane in a non-polar form and then become impermeable following hydrolysis by cellular esterases (Rogers *et al.* 1983). Glass calcium-selective microelectrodes filled with highly specific ion-exchangers are also available now (Coray *et al.* 1980; Baker and Umbach 1983). However, they cannot reach the level of sensitivity given by chemical probes.

State of calcium in the neuron

The first measurements of free (ionized) calcium with aequorin were made on squid axons and gave a value of 0.3 μM (Baker *et al.* 1971). During stimulation this level increased, and EGTA injection abolished the response. However, the total concentration of calcium in the axons during these experiments was extremely high, probably due to long storage after isolation; therefore, the level of ionized calcium could also be elevated. In fact, in experiments carried out on freshly isolated axons, DiPolo *et al.* (1976) estimated the level to be about 0.02 μM. Using arsenazo III, the same authors found somewhat higher values (0.05 μM); as such levels are near the limits of arsenazo sensitivity, this determination could not be very precise. In any case, the amount of free Ca^{2+} in the axoplasm is extremely small, only about 60 ions in μm^3 (Baker 1976).

Measurements on other cells gave quite variable results. In different muscle fibres during rest the level of ionized calcium was found to be: in sheep heart Purkinje fibres, 0.17 μM, and ventricular fibres, 0.36 μM; and in frog skeletal muscle fibres, 0.05 μM (Coray *et al.* 1980). In giant snail neurons glass-microelectrode measurements gave a level of 0.17 μM (Alvarez-Leefmans *et al.* 1981). The use of arsenazo III showed lower values ($\cong$0.01 μM); during stimulation it increased by about five times (Gorman and Thomas 1978; Gorman *et al.* 1981). In our experiments, detailed measurements of ionized calcium and its changes during different treatments of *Helix pomatia* neurons were carried out using the intracellular indicator Fura-2 (Tepikin *et al.* 1987; Kostyuk *et al.* 1989*a*). No compartmentation of the indicator, often seen in other cells, has been observed; this may be due to the relatively low temperature used during measurements (about 20 °C). Probably, at higher temperatures, which activate the accumulation of calcium ions in intracellular organelles, a spatial heterogeneity would be more obvious. Direct injection of the non-esterified form of Fura-2 instead of extracellular application of Fura-2 AM gave identical results, as well as fluorescence measurements using either a single- or double-wave technique (Grynkiewicz *et al.* 1985).

The basal level of ionized calcium was found to be 0.09 ± 0.07 μM (38 cells) with variations between 0.07 and 0.15 μM. This is much less than the values obtained with calcium microelectrodes and closer to the data from arsenazo measurements.

Binding and sequestering of calcium in the cell

The marked ability of the cell to bind Ca^{2+} is due to the presence in it of several effective buffer systems. First of all, these systems include cytoplasmic proteins, some of which bind Ca^{2+} very specifically (for instance, calmodulin). Protein binding has a high rate (in the millisecond range) but low capacity (2–3 nM mg^{-1} protein), see Blaustein (1988). Therefore, it can comprise only a small part of the total buffering capacity of the cytoplasm, and of greater importance must be the accumulation of Ca^{2+} by different intracellular organelles. Obviously, the latter type of buffering will vary in different parts of the cell because of the uneven distribution of the organelles. Such inequality has actually been demonstrated on nerve cells both photometrically and roentgenoscopically. X-ray microanalysis of mollusc neuronal sections has shown that the calcium content in some parts (axon hillock) is about twice as high as that in other parts (Sugaya and Onozuka 1978). The distribution of some cytoplasmic structures, among them the endoplasmic reticulum in the axon hillock, is also very different from the rest of the cell soma. In physiological experiments on the same neurons, local injections of $CaCl_2$ from a microelectrode into different parts of the cell soma previously loaded with arsenazo III were performed. The largest buffering capacity was detected near the surface membrane; the lowest, near the nucleus. The thickness of the near-membrane sheath with increased buffering was about 50 μm (Tillotson and Gorman 1980).

Obvious candidates for calcium accumulators are the mitochondria, which have special ion transport mechanisms. More detail concerning the mechanism of calcium transport in the mitochondrial membrane can be obtained from reviews (for example, Fiskum and Lehninger 1980). The basis for such transport is formed by electrophoretic transfer of Ca^{2+} through the inner membrane, due to the existence on this membrane of a considerable transmembrane potential (the inner side of the membrane is negative with respect to the outer one). This potential difference is formed mainly by the transfer of electrons across the membrane during respiration. The influx of each Ca^{2+} ion is compensated by the transport of two protons in the opposite direction, inducing an alkalinization of the interior space of the mitochondria. An excessive increase in pH is prevented by a parallel movement of protons towards the inside, coupled to the transport of phosphates.

It should be noted that even after blocking respiration (for instance by cyanides), the mitochondria still retain the ability to accumulate calcium. This process depends on the presence of ATP because ATP itself supports the membrane potential on the inner mitochondrial membrane by transporting protons to the outside. A decrease or removal of this potential correspondingly affects the accumulation of Ca^{2+}. Such an effect can be produced by potentiation of the inward flux of protons; all agents that increase this flux decrease, in parallel, the accumulation of calcium.

The immediate carrier for Ca^{2+} seems to be a special protein which has two calcium binding sites (judging from the Hill coefficient for such transport, which is about 2). Attempts to isolate this protein resulted in the separation of several glycoproteins which bind Ca^{2+} with high affinity. Antibodies against some of them inhibit calcium transport in mitochondria (Panfili *et al.* 1976). From the inner membranes a small hydrophobic peptide (molecular weight about 3000 Da) has also been isolated, which binds Ca^{2+} with $K_d = 5.2\ \mu M$ (Jeng and Shamoo 1980). The properties of this peptide, designated calciforin, are close to those of a suggested electrophoretic calcium-carrier; the translocation of calcium induced by this peptide could be inhibited by ruthenium red, a mitochondrial blocker.

Calcium ions accumulated in the mitochondria can be released again. This can be done partly by the reversal of the carrier mechanism. However, different mitochondrial poisons inhibit the release only partially. This may indicate the presence of the Na/Ca exchange mechanism (see below).

The suggestion that mitochondria play a role in the regulation of Ca^{2+} level in the nerve cell has been tested experimentally. In experiments with the axoplasm extruded from giant axons, addition of cyanide or carbonyl cyanide *p*-trifluoro-methoxy-phenylhydrazone (FCCP) increased the level of ionized calcium by 2–3 times (Baker *et al.* 1971; Mullins and Requena 1979). However, as shown by Brinley *et al.* (1977), the amount of released ions was only a small proportion of the bound ions (not more than 10 per cent). In our experiments on snail neurons with the use of Fura-2, application of the mitochondrial poisons sodium azide or carbonyl cyanide phenylhydrazone (CCCP) increased the intracellular Ca^{2+} level to about 0.3 μM; this effect developed and recovered slowly (many tens of seconds), as can be seen from the example shown in Fig. 8.1. After the inhibition of mitochondrial transport by these agents, the cell retains its ability to maintain a physiologically low level of ionized calcium (Requena *et al.* 1977). Evidence of a minor role of mitochondria in the accumulation of intracellular Ca^{2+} is also seen from

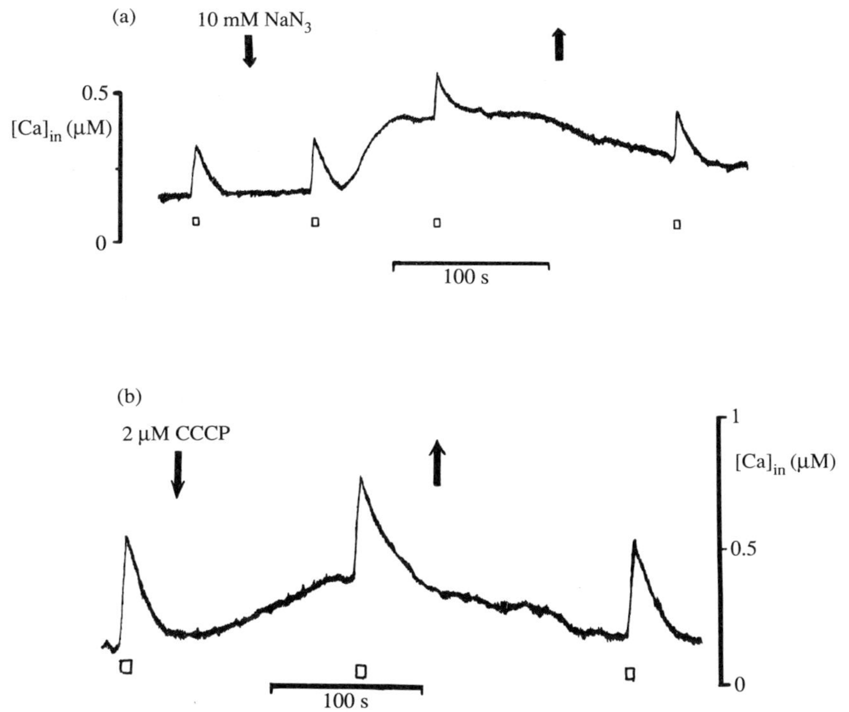

Fig. 8.1. Changes of $[Ca^{2+}]_{in}$ level in a snail neuron after application of mitochondrial poisons (Kostyuk *et al.* 1989*a*). (a) 1 mM sodium azide; (b) 2 μM CCCP. Marks indicate periods of membrane depolarizations to 0 mV.

studies of the concentration dependence of this process (Baker 1976). The uptake of Ca^{2+} ions by both isolated mitochondria and *in vivo* reached a half-maximal value at concentrations of about 20–40 μM and developed very slowly; thus, it is a low-affinity process (compared to physiological intracellular concentrations). Obviously, the bulk of intracellular calcium is sequestrated by some other more sensitive mechanism(s) which may have a lower capacity but can buffer these ions rapidly at very low concentrations; thus, such mechanisms may play a key role in the rapid restoration of the basic level of Ca^{2+} after periods of cellular activity connected to the opening of calcium channels. Most probably, these mechanisms are found in the endoplasmic reticulum (Requena and Mullins 1979). The special cisternae localized immediately beneath the cell membrane—the so-called calciosomes—may be of great importance (Meldolesi *et al.* 1988). An example of such non-mitochondrial accumulation was first obtained on brain synaptosomes in which, after the combined action of all possible factors inhibiting the mitochondrial transport, a considerable ATP-dependent calcium accumulation still remained (Blaustein *et al.* 1978*a*, *b*). This component could be characterized by high affinity (half saturation at $[Ca^{2+}]_{in}$ about 0.35 μM) but low capacity. The mitochondrial accumulation in this case has a much lower affinity (above 10 μM) but a high capacity (Blaustein *et al.* 1978*b*). Later, the important role of the endoplasmic reticulum in the accumulation of calcium in nerve cells under physiological conditions was shown in our measurements of the relaxation of calcium transients induced by activation of calcium channels (see below).

The molecular mechanism of calcium accumulation in the endoplasmic reticulum has not yet been studied directly; it may well be related to a similar mechanism in the sarcoplasmic reticulum of muscle fibres, where it is caused by a very effective ATP-driven calcium pump. The activity of this pump is subjected to complex regulation by cellular metabolic processes; in cardiac fibres an important role in such regulation is played by an integral peptide, phospholamban (molecular weight 22 kDa), which can be phosphorylated by both cAMP-dependent and Ca-calmodulin-dependent protein kinases (PK-A and Calm-PK) (see Carafoli and Zurini 1982; Carafoli 1988). In skeletal muscle fibres another glycoprotein was detected (molecular weight 53 kDa) which is tightly connected to the reticulum membranes and possibly participates in the regulation of the activity of the corresponding calcium pump (Chiesi and Carafoli 1982). A similar analysis of the calcium pump in the neuronal endoplasmic reticulum would be very important.

Sodium–calcium exchange across the neuronal membrane

Despite the effectiveness of intracellular calcium binding and accumulating systems, their capacity is not unlimited, and during natural activity of the neuron they would become saturated. Therefore, in parallel cellular mechanisms must exist which would extrude the excess Ca^{2+} back into the extracellular medium. One such mechanism is that of sodium–calcium (Na/Ca) exchange.

The existence of Na/Ca exchange was first shown in cardiac muscle fibres on the basis of the dependence between ^{45}Ca extrusion and the concentration of Na^+ in the extracellular medium. A decrease in $[Na^+]_{out}$ depressed the outflow of Ca^{2+}, indicating the dependence of the latter on the energy of the sodium electrochemical gradient (Lüttgau and Niedergerke 1958; Reuter and Seitz 1968). This suggestion was later extended to the axonal membrane, in which the efflux of ^{45}Ca is also dependent on the Na^+ influx (Blaustein and Hodgkin 1969; DiPolo 1973). After replacement of extracellular Na^+ by Li^+, choline, or dextrose, the efflux of ^{45}Ca became depressed to about 50 per cent of its initial level. The remaining component depended on the presence of extracellular Mg^{2+} ions and decreased to only 10 per cent of the initial value after their removal. For quantitative analysis and elimination of the possible liberation of Ca^{2+} from internal stores, the experiments were repeated in conditions of stabilized intra-axonal ionized calcium by injection of calcium buffers; in these conditions the activation of ^{45}Ca efflux was still similar to that in axons without calcium-buffer injection (Baker and McNaugton 1976). In axons poisoned by cyanide, the replacement of extracellular Na^+ by Li^+ also induced a strong depression of ^{45}Ca efflux (Blaustein and Hodgkin 1969). Confirmation of the role of Na/Ca exchange in the somatic membrane for the maintenance of the stationary level of intracellular Ca^{2+} has been obtained by Kostyuk *et al.* (1989*a*), working on snail neurons. After transfer into sodium-free solution (thereby reversing the direction of the exchange mechanism) the level of $[Ca^{2+}]_{in}$ increased slowly to 0.1–0.2 μM; after readmission of sodium it returned to initial values (Fig. 8.2).

Clearly, the energy of sodium electrochemical gradient is sufficient to drive the opposite transfer of calcium ions; the energy of ATP is not necessary here. However, ATP has some catalytic effect, probably by increasing the affinity of the transporting mechanism for Ca^{2+} without altering the exchange rate (see Blaustein 1976). The Na/Ca exchange in the sarcolemma of cardiomyocytes can also be activated by ATP, Mg^{2+}, and Ca^{2+}, and deactivated by dephosphorylation. Activation can be inhibited by calmodulin-antagonists, indicating the possible participation of Ca-calmodulin-dependent protein

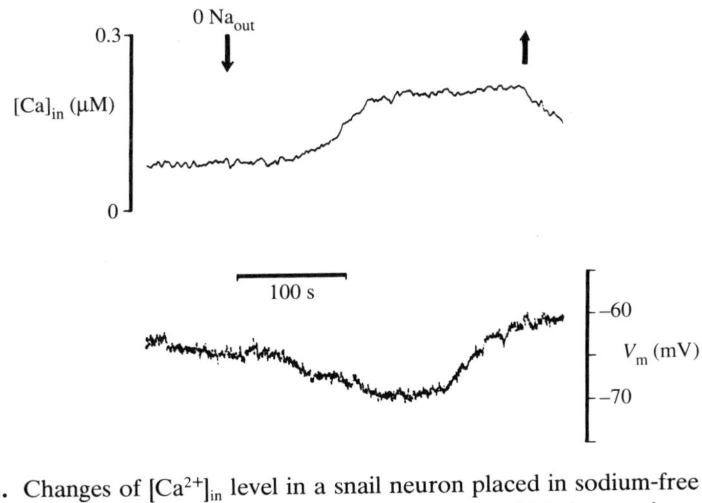

Fig. 8.2. Changes of $[Ca^{2+}]_{in}$ level in a snail neuron placed in sodium-free solution (Kostyuk *et al.* 1989*a*). The lower record presents parallel changes in membrane potential.

kinase (Calm-PK) in the function of the Na/Ca exchanger (Caroni and Carafoli 1983).

Attempts to determine the sensitivity of the exchange system gave somewhat contradictory results. For the axonal membrane the apparent dissociation constant for Ca^{2+} ions seems to be 50–100 times above the physiological level of free calcium. In this case it would be difficult to explain the physiological data indicating considerable participation of the exchange mechanism in the maintenance of the low level of intracellular ionized calcium; this mechanism should be classified as one with high capacitance but low affinity (DiPolo and Beaugé 1983). Similar findings of a low affinity were obtained for the Na/Ca exchange system of heart sarcolemmal vesicles: $K_d^{Na} = 16$ mM and $K_d^{Ca} = 1.5\ \mu$M; these values are also above the physiological level of free calcium in the fibres (Caroni *et al.* 1980).

It is clear that it will be important to find additional ways of identifying the putative carriers that provide the Na/Ca exchange in the neuronal membrane. Pharmacological approaches are not very promising in this respect. It is known that the action of other divalent cations on the exchange mechanism differs from that on calcium channels; for instance, the former can be inhibited by Sr^{2+} ions which readily pass through calcium channels (Blaustein 1977). However, the difference is not sufficient for a reliable separation of both mechanisms. In contrast to Ca-dependent-ATPase (see below), the Na/Ca exchange, at least in cardiac cells, can be inhibited by chlorpromazine (50 per cent inhibition at 90 μM), but it is not sensitive to vanadate which blocks many transport ATPases (Caroni *et al.* 1980).

The stoichiometry of the Na/Ca exchange is also quite important for understanding its functional role. In the initial experiments on cardiac muscle fibres (Reuter and Seitz 1968) it was estimated as 2 Na:1 Ca (electroneutral). But later, in experiments on giant axons, the relations 3 Na:1 Ca or even 4 Na:1 Ca were found (Mullins 1977); in this case the exchange mechanism should separate charges and generate transmembrane currents. Moreover, in physiological conditions (during an action potential) a reversal of exchange may take place. For instance, if the stoichiometry is 4 Na:1 Ca, the reversal potential for ion exchange must be about 50 mV ($E_{ex} = 2E_{Na} - E_{Ca}$; $E_{Na} \cong +40$ mV, $E_{Ca} \cong +130$ mV). Therefore, during membrane potential displacement to more positive values a stimulation of Na^+ efflux and of Ca^{2+} influx should take place; during repolarization the exchange will again acquire its usual direction (see Fischmeister and Vassort 1981).

Calcium pump

As already mentioned, the experiments on squid giant axons have shown that a considerable part of ^{45}Ca efflux is not dependent on the sodium gradient;

this component is better expressed in axons which maintain a low physiological level of ionized calcium and depends on the presense of ATP (DiPolo 1977). It is important that the ATP-dependent Ca^{2+} efflux takes place in the absence of any other ionic gradient; thus, it represents an uncoupled active transport (i.e. not connected with simultaneous transfer of other ions into the cell).

The most detailed investigations of the active transport of calcium were performed on erythrocyte and cardiomyocyte membranes, where calcium is carried by a special membrane enzyme, Ca-dependent ATPase (reviewed by Carafoli and Zurini 1982). Its specific activation by calmodulin was used for separation by affinity; for erythrocytes a protein with a molecular weight of 138 kDa was obtained which hydrolyses ATP in a highly Ca-dependent manner. After proteolytic fragmentation several polypeptides were obtained with molecular weights ranging from 14 to 124 kDa. The sensitivity to calmodulin disappeared together with the 90 kDa fragment, which probably carries the calmodulin receptor. This fragment loses the ability to bind calmodulin if a particle with molecular weight 8 kDa is cleaved from it; the latter may be the calmodulin receptor.

Being incorporated into liposomes, the intact protein transported Ca^{2+} with a stoichiometry 1 Ca^{2+}:1 ATP. Such uncoupled transport must be electrogenic; the potential produced is probably neutralized by the passive transmembrane movement of anions. From cardiomyocytes a somewhat larger molecular complex (150 kDa) has been isolated with similar transport properties; it could be differentiated from the ATPase of the sarcoplasmic reticulum by its high sensitivity to vandadate. A cAMP-linked phosphorylation process stimulated the ATPase and the related pumping of calcium.

The most important property of the Ca-dependent ATPases shown in these experiments is their very high affinity for Ca^{2+}. The apparent K_d for the ATPases from cardiomyocytes was found to be about 0.3 μM, i.e. close to the physiological level of intracellular ionized calcium. Therefore the calcium pump rather than Na/Ca exchange may play the main role in the maintenance of this low calcium level (although the rate of ion transfer by the pump is much lower than by the exchange mechanism).

This conclusion can be extended to the neuronal membrane: here the calcium pump also represents a high-affinity transport system with low capacity. Its 50 per cent activation can be achieved by Ca^{2+} concentrations only twice the physiological level in the squid giant axon, which is 100 times less than required for the Na/Ca exchange system. Biochemical investigations of the calcium pump in neuronal membranes are still lagging behind those in other membranes. Studies of membrane fragments from squid optic nerve have shown a high degree of similar Ca-dependent ATPase activity (50 per cent activation at 0.3 μM Ca^{2+} and 18 μM ATP, almost complete inhibition by micromolar concentrations of vanadate). Mg^{2+} ions were necessary for the activation of this ATPase (DiPolo and Beaugé 1983).

Obviously, more extensive investigations of the calcium pump in other neuronal membranes would be highly desirable.

The findings about the biochemical mechanisms of the Na/Ca exchange and calcium pump raise some interesting suggestions. As was mentioned, both mechanisms need for their function the presence of ATP (although in the pump it is hydrolysed and in the exchange mechanism it only changes the affinity of the carrier to Ca^{2+}). Both mechanisms are subjected to phosphorylation by protein kinases, and the phosphorylated state is the active one; dephosphorylation leads to deactivation of ion transport. Perhaps both systems are not quite independent from the structural point of view and represent different functional modes of the same pumping mechanism, which starts to act as an ionophore at high Ca^{2+} levels near the inner surface of the membrane.

Calcium signals during neuronal excitation

Despite the presence of all the described systems, which may stabilize the intracellular level of ionized calcium, a considerable increase in $[Ca^{2+}]_{in}$ takes place in the nerve cell during excitation, which can be described as the intracellular *calcium signal*. The changes of $[Ca^{2+}]_{in}$ during cellular activity are very different in their kinetic characteristics from those appearing during treatments that interfere with the calcium transporting systems. Such signals have been investigated by our group on snail neurons, using the fluorescent probe Fura-2 (Tepikin *et al.* 1987; Kostyuk *et al.* 1989*a*), and by other groups.

During membrane depolarization to levels sufficient for spike generation and opening of high-threshold calcium channels, a rapid increase of $[Ca^{2+}]_{in}$ takes place. (As was already mentioned, the low-threshold calcium channels are poorly represented in snail neurons and can be neglected in the analysis.) The rate of rise depended on the frequency of stimulation and external calcium concentration, as can be seen from recordings presented in Fig. 8.3a. After termination of stimulation, $[Ca^{2+}]_{in}$ returned to a background level with a relatively constant time course (5–30 s in different neurons, mean value 20 ± 5 s, $n = 27$). The recovery rate was highly temperature-dependent, decreasing with cooling ($Q_{10} = 2.3 \pm 0.2$). Such dependence illustrated in Fig. 8.3b indicates a major role for metabolic processes in the termination of the signal. To determine the possible mechanisms involved, the calcium transients evoked by identical stimuli were compared before and after reversal of the Na/Ca exchange, poisoning of the mitochondria, and addition of 1 mM La^{3+} or vanadate to the external solution; none of these factors induced any detectable changes in the recovery rate, as can be seen from the illustrations. However, a substantial change appeared after injection of vanadate into the neuron, which depresses the calcium pump in the

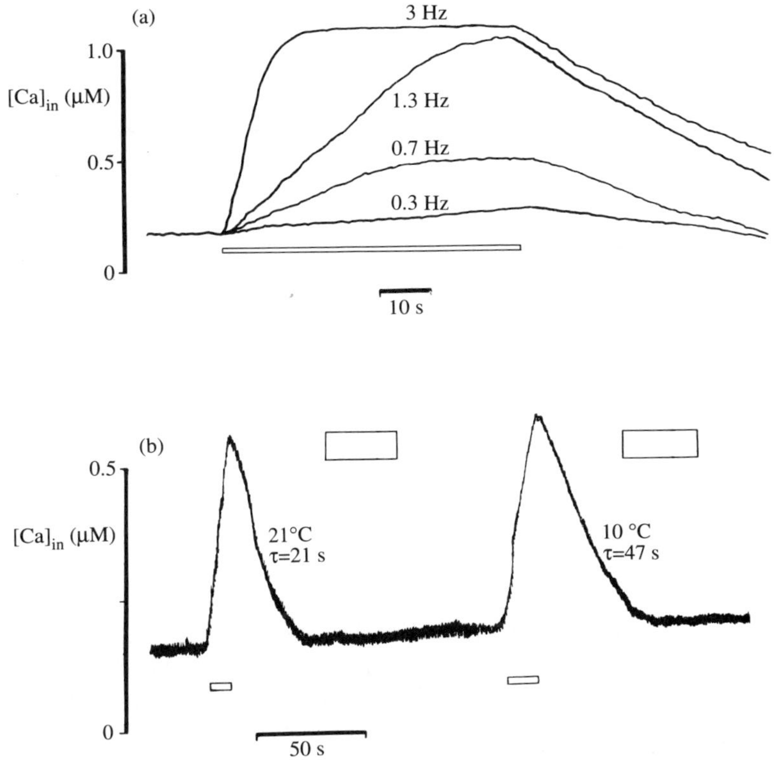

Fig. 8.3. Development and recovery of $[Ca^{2+}]_{in}$ changes induced by membrane depolarization in a snail neuron (Kostyuk *et al.* 1989*a*). (a) Rhythmic stimulation by pulses (100 ms long) of different frequency (frequencies are indicated near the record). (b) Temperature dependence of recovery (time constants are indicated near the records).

membrane of intracellular organelles (endoplasmic reticulum). Such an injection led to a progressive irreversible increase of $[Ca^{2+}]_{in}$, illustrated in Fig. 8.4.

These findings are in accordance with the idea that the main mechanism by which the calcium entering the cell following the activation of voltage-operated calcium channels is stored is the active uptake of ions by the endoplasmic reticulum, due to activation of the calcium pump in its membrane—as happens in muscle fibres due to the activity of the sarcoplasmic reticulum. It should be mentioned that Miledi *et al.* (1982), in their study on calcium transients in skeletal muscle fibres with the use of arzenazo III, also found a $Q_{10} = 2.4$ for the relaxation rates. However, in muscle fibres

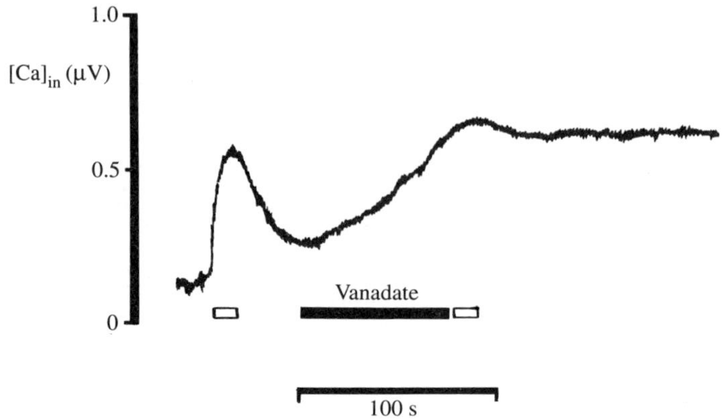

Fig. 8.4. Changes in $[Ca^{2+}]_{in}$ after intracellular injection of vanadate into a snail neuron (Kostyuk *et al.* 1989*a*). The duration of injection is shown by the shaded bar, the depolarizations to 0 mV by unshaded bars.

the absolute value of the relaxation rate was about two orders of magnitude higher than that found in nerve cells. Ahmed and Connor (1988) obtained a relaxation time constant of about 7 s in neurons from the mollusc *Anisodoris*; Gorman and Thomas (1978) and Gorman *et al.* (1984) reported values of between 5 and 20 s in *Aplysia* neurons.

Measurements of calcium transients can be also used for the determination of the calcium buffering capacity of the nerve cell. For this purpose, Ahmed and Connor (1988) used injections into a snail neuron of measured quantities of EGTA, which diminish the calcium signal; the capacity value was 45.2 $\mu M/\Delta pCa$.

A characteristic change in the kinetics of the calcium signal appeared when 5 mM caffeine was added to the extracellular solution; when the evoked response reached a threshold level of 0.4 ± 0.05 μM, additional spontaneous elevation of $[Ca^{2+}]$ occurred in an 'all or none' manner. At negative membrane potentials, more negative than -50 mV, it was accompanied by a distinct membrane hyperpolarization, probably due to activation of a calcium-dependent potassium conductance. An example of this effect is presented in Fig. 8.5. Similar results could also be obtained by 5 mM theophylline, but in this case the necessary threshold elevation of Ca^{2+} was higher (0.6 ± 0.05 μM). The maximum amplitudes of additional elevation reached 1 μM and did not depend on the concentration of $Ca^{2+}{}_{out}$ in the range between 0.5 and 20 mM. When the threshold concentration of caffeine was about 2 mM, the decay rate of additional elevation was similar to that of the evoked one.

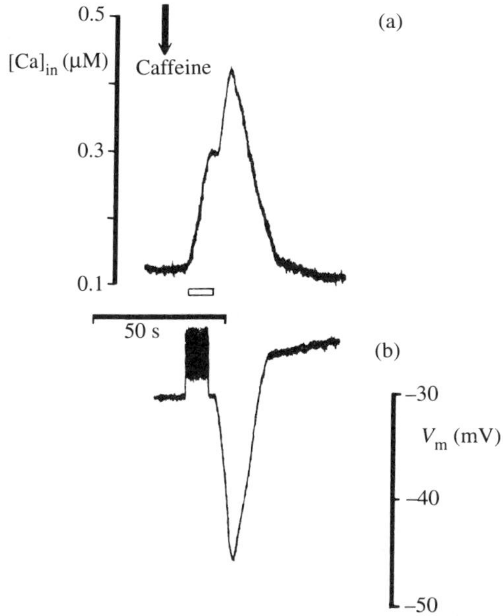

Fig. 8.5. (a) Calcium release in a snail neuron triggered by depolarization-induced Ca-load in the presence of 5 mM caffeine (Kostyuk *et al.* 1989*a*). (b) Parallel changes in membrane potential. The duration of depolarization to 0 mV is indicated by the unshaded bar.

In this case a caffeine-dependent calcium-induced release of Ca^{2+} from the intracellular stores must take place, as occurs in muscle fibres from the sarcoplasmic reticulum. A similar conclusion was made by Thayer *et al.* (1988) from a study of calcium signals in rat sympathetic neurons. In muscle fibres such a release of intracellular Ca^{2+} can be blocked by some local anaesthetics. In our experiments on snail neurons, lidocaine (1 mM) did not influence the induced release, but the same concentrations of procain and tetracaine gradually depressed it. In sympathetic neurons the release could be disturbed by specific modifiers of the sarcoplasmic reticulum release channels, ryanodine and dantrolen.

In many neurons spontaneous oscillations of $[Ca^{2+}]_{in}$ appeared in the presence of caffeine or theophylline, which could continue in a steady rhythm for hours (see Fig. 8.6). The waves of elevated calcium could be accompanied by synchronous waves of membrane hyperpolarization. Usually, initial depolarization leading to the opening of calcium channels and elevation of the intracellular calcium level was necessary to trigger the oscillations; their frequency could be modulated by changing the calcium influx. Calcium-release blockers disturbed the oscillatory rhythm.

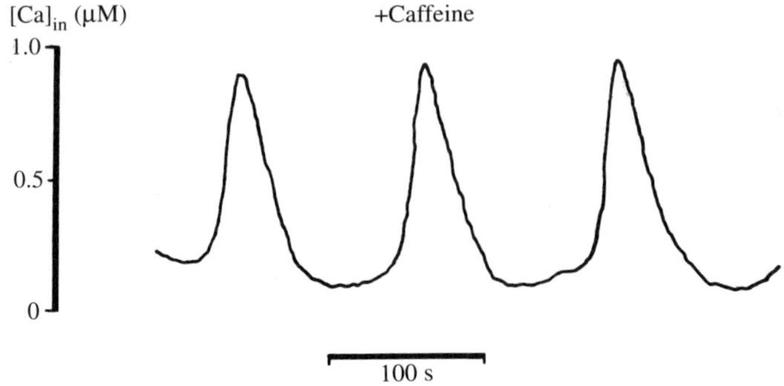

Fig. 8.6. Cyclic changes in $[Ca^{2+}]_{in}$ observed in a snail neuron in the presence of caffeine (Kostyuk *et al.* 1989*a*).

Similar oscillations of $[Ca^{2+}]_{in}$ have been described in frog sympathetic neurons, where they too were related to calcium release from caffeine-sensitive calcium stores (Lipscombe *et al.* 1988*a*). They also occur in cells from other tissues, for instance in parotid salivary cells (Gray 1988) and somatotrophic cells of the anterior pituitary (Holl *et al.* 1988); the nerve cells differed only in the necessity to add caffeine to trigger the oscillations. Such oscillations can be predicted theoretically on the basis of interaction between release and accumulation processes in the endoplasmic reticulum; a necessary condition to start them is the initial activation of calcium conductance in the cell membrane (Mironov and Tepikin 1988). They may be of importance in the genesis of caffeine-induced seizure activity of the brain.

As the caffeine-dependent calcium release in the nerve cell is very similar to calcium release from the sarcoplasmic reticulum, we might suggest that it is exerted through analogous release channels in the membrane of the endoplasmic reticulum. It is known that sarcoplasmic reticulum release channels are formed by a protein of molecular weight about 170 kDa, which easily aggregates into larger polymeric complexes (Imagawa *et al.* 1987*b*; Fill and Coronado 1988; Rubtsov and Murphy 1988; Anderson *et al.* 1989). The purification of this protein became possible due to its specific binding to ryanodine, which opens the corresponding channels and, in this way, depletes the calcium stores. The protein can be incorporated into a lipid bilayer where it forms channels with high unitary conductance, often disintegrating into sublevels; the channels are quite selective for divalent cations, with P_{Ca} or $P_{Ba}/P_{Tris} = 8.7$. The reconstructed channels can be activated by Ca^{2+} in the presence of ATP, and blocked by Mg^{2+} and ruthenium red (Smith *et al.* 1988; Imagawa *et al.* 1987*b*; Fill and Coronado 1988; Hymel *et al.* 1988; Lai *et al.* 1988). All these findings can only be used as a starting point for the description of calcium-release channels in nerve cells.

Our understanding of the mechanisms forming the calcium signal in the nerve cell recently became complicated by the finding that calcium depots may exist here which are regulated in a different way—by inositol 1,4,5-triphosphate (IP_3). The IP_3-induced release of Ca^{2+} is known in muscle fibres. It requires the presence of GTP and is possibly mediated by GTP-binding proteins (Saida and van Breemen 1987). In muscle fibres it is much less marked than the calcium-induced release and probably does not play significant role in excitation–contraction coupling (Fabiato 1988).

However, inositol phosphates can also induce substantial calcium release in neurons, as shown by Belan *et al.* (1989), using *Helix pomatia*, and illustrated in Fig. 8.7. IP_3 was injected under pressure from one barrel of a double-barrel microelectrode; the second barrel was used for potential recording or stimulation. The concentration of IP_3 achieved in the cell was

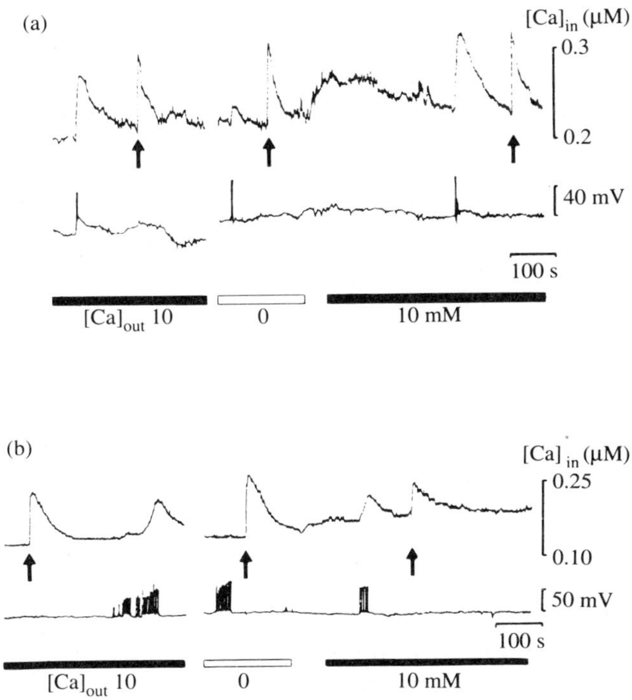

Fig. 8.7. Calcium transients evoked in snail neurons by injection of IP_3 (a) and its non-hydrolisable analogue Gpp[NH]p (b) in the presence and absence of $[Ca^{2+}]_{out}$ (Belan *et al.* 1989). Arrows indicate identical injections; shaded and unshaded bars show the changes in external solution. The lower records show changes in membrane potentials.

about 10 μM, and that of the non-hydrolysed GTP analogue Gpp[NH]p was about 100 μM. Following IP_3 injection, the intracellular level of Ca^{2+} increased by about 0.1 μM, the increase continuing after the end of injection. The effect did not change after substitution of the normal extracellular solution by a calcium-free one. A similar effect could be produced by the injection of Gpp[NH]p. The high speed of calcium release and the absence of a detectable delay indicate a possible direct action of GTP and its analogues on the structures responsible for calcium release, perhaps via GTP-binding proteins. A calculation has shown that one IP_3 molecule can induce the liberation of about 20 calcium ions (see Osborn *et al.* 1988). A similar liberation of Ca^{2+} after injection of IP_3 has been shown in *Aplysia* bag cells; it occurred mainly in the cell soma and was followed by the activation of potassium channels in the cell membrane (Fink *et al.* 1988). Considerable IP_3-induced liberation of Ca^{2+} has been found in *Xenopus* oocytes, where it triggers rhythmic oscillations of $[Ca^{2+}]_{in}$ similar to those described in neurons (Berridge 1988). Especially effective IP_3-induced liberation of calcium occurs in lymphocytes (Meyer *et al.* 1988).

The natural agonist which induces the accumulation of IP_3 and subsequent liberation of calcium in a nerve cell is not yet known. It has been shown on clonal phaeochromocytoma cells that these processes can be evoked by bradykinin; they lead to the opening of membrane channels and additional influx of ions into the cell (Fasolato *et al.* 1988). The same happens in cultured chromaffin cells after the external application of ATP (30 μM), which probably acts via purine receptors (Sasakawa *et al.* 1989). In clonal GH_3 pituitary cells an elevation of $[Ca^{2+}]_{in}$ remaining in calcium-free external solution can be induced by the action of thyrotropin releasing hormone (TRH) (Benham 1989).

One should keep in mind one more possible mechanism of IP_3 action–modulation of voltage-operated calcium channels in the cell membrane. Vilven and Coronado (1988) have shown that IP_3 in micromolar concentrations can open DHP-sensitive channels from the T-system of skeletal muscle fibres incorporated in a lipid bilayer, acting like their intracellular agonist.

Conclusions

The available experimental data indicate that the extremely low level of ionized calcium characteristic for the interior of nerve cells (10^{-8}–10^{-7} M) is maintained by a variety of cellular mechanisms (Fig. 8.8). On the one hand, the cell possesses systems for the effective accumulation of Ca^{2+}, represented mainly by mitochondria and endoplasmic reticulum. These structures accumulate calcium ions by transporting them actively through their membranes. The sensitivity of the endoplasmic reticulum to calcium is higher than that of mitochondria; for this reason the former probably plays a

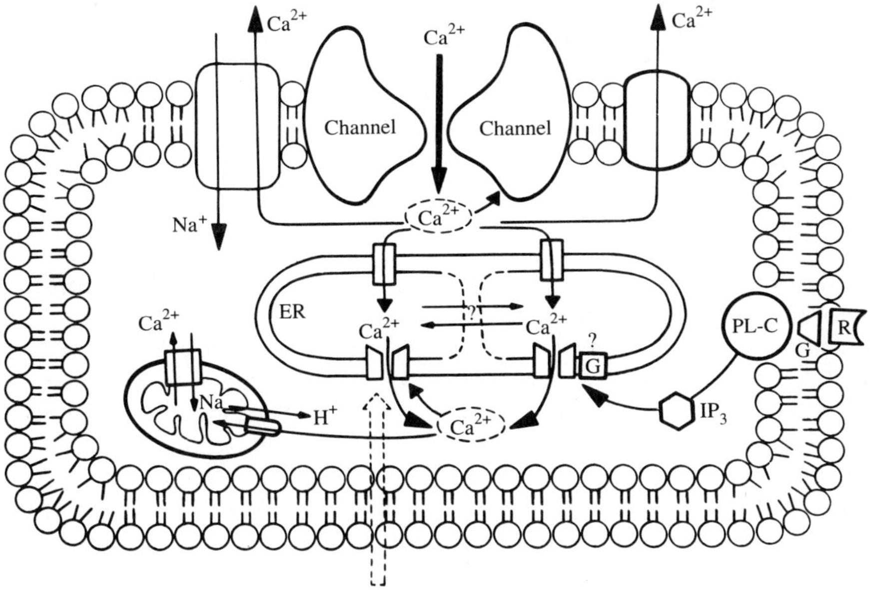

Fig. 8.8. Schematic presentation of the intracellular mechanisms participating in the regulation of free calcium in a nerve cell. ER, endoplasmic reticulum, possibly separated into Ca–caffeine- and IP_3-sensitive pools. PL-C, phospholipase C; IP_3, inositol-1,4,5-triphosphate.

dominant role in the fast restoration of the low $[Ca^{2+}]_{in}$ after elevation during cellular activity. Calcium ions accumulated in the reticulum can also be rapidly released during activity due to the opening of specific calcium channels in the membrane. These channels can be opened in the nerve cell by elevation of $[Ca^{2+}]_{in}$ and by the formation of inositol 1,4,5-triphosphate.

On the other hand, the neuronal membrane has two systems for the extrusion of excessive calcium from the cell: Na/Ca exchange and the calcium pump. The first uses the energy of the sodium transmembrane gradient and the second the energy of ATP. The calcium pump is activated at lower intracellular calcium levels than the Na/Ca exchange and can serve, therefore, as the main mechanism for the maintenance of physiologically very low $[Ca^{2+}]_{in}$. On the contrary, the exchange system has a high capacity and can produce considerable transmembrane ion fluxes. The Na/Ca exchange is not neutral and therefore may be depend on membrane potential; under certain conditions it can reverse its direction and serve as an additional way for injection of Ca^{2+} in the cell.

9. Functional role of calcium ions in the nerve cell

History

The functional role of the influx of Ca^{2+} ions from the extracellular medium into the cell during its activity is not always obvious. In some structures such an influx is directly connected to the triggering of a specific cellular function, and the role of these ions as messengers between the cellular membrane and the effector apparatus can be precisely analysed. This is the case for synaptic terminals which liberate a transmitter, and muscle fibres which contract in response to membrane depolarization. The important role of calcium in muscle contraction had already been defined by Ringer at the end of the nineteenth century. Subsequent numerous investigations have firmly established the fact that in all kinds of muscle fibres the intracellular level of ionized calcium is the primary regulator of the contraction–relaxation cycle, although some details of such regulation, particularly the role of Ca^{2+} ions entering from the extracellular space and liberated from intracellular stores, may substantially differ in different fibres.

The messenger role of Ca^{2+} in the triggering of transmitter release from nerve endings was detected much later, in the classical experiments of Katz and Miledi (1967*a*, *b*) on neuromuscular synapses. The results of these experiments were so striking that they immediately formed the basis of modern understanding of the mechanism of synaptic transmission, although several details of the mode of action remain to be resolved.

However, an effective mechanism for the injection of calcium through the cell membrane is also present in structures having no obvious effector functions; these are the neuronal soma and dendrites. The density of voltage-operated calcium channels in the somatic membrane of many neurons is so high that the inward current produced by their activation can be strong enough to generate a regenerative response in the form of an action potential. In some cells the calcium component of the inward current exceeds the sodium one, and even at normal concentrations of electrolytes in the extracellular medium the action potential is predominantly a 'calcium' spike. The question, 'Why are calcium channels so effectively used for the generation of the nerve impulse in the somatic (and dendritic) membranes?' has attracted much attention. First of all, the slowness of activation, and especially inactivation, of calcium channels has to be taken into account. The prolonged time course of the calcium currents and of the corresponding membrane potential changes can be quite important in the regulation of the rhythmic activity of the nerve cell; noteworthy in this respect was the discovery that calcium ions entering the cell in turn activate outward (potassium) currents (Meech 1974). This aspect of the problem has also been analysed by modelling the electric phenomena occurring in dendritic and somatic membranes (Llinas and Sugimori 1980*a*, *b*; Gorman *et al.* 1981; Mironov 1983*a*); such analysis has shown qualitatively that 'slow' calcium currents can be important in the determination of the discharge pattern of the cell.

On the other hand, more and more data have been accumulated indicating that Ca^{2+} influx can play a more general role in the functioning of a nerve cell. Investigations concerning this problem became more significant after the detection of cyclic nucleotide metabolism and the important role of Ca^{2+} in its performance (Rasmussen *et al.* 1979). As phosphorylation of different proteins by cyclic-nucleotide-dependent kinases appeared to be an effective way for the regulation of cellular processes, the possible participation of calcium ions in such regulation was recognized.

Regulation of the ionic permeability of the neuronal membrane

The recurrent action of Ca^{2+} ions entering the cell through the surface membrane on the same membrane is obviously their fastest functional effect and it will be considered first.

As already described in Chapter 4, elevation of $[Ca^{2+}]_{in}$ produced by the activation of high-threshold calcium channels inactivates the same channels. In parallel, special potassium channels become activated. This effect has been detected by direct injection of calcium in snail neurons through a microelectrode; the injection was followed by membrane hyperpolarization and a decrease in membrane resistance. The activation of a potassium current has been proved on the same neurons in voltage-clamp experiments (Meech and Standen 1975). An N-shaped inflection has been observed in the positive region of the current–voltage characteristic of the outward current, which disappeared after placing the cell in a calcium-free extracelluar solution. This additional calcium-dependent component of the outward current was interpreted as the result of activation of special potassium channels by the recurrent action of Ca^{2+} ions entering the cell through voltage-operated calcium channels. At very strong depolarizations this component decreased in amplitude and finally disappeared, clearly because of the decrease in the inward calcium current. The phenomenon appears to be widespread: it was also observed in smooth muscle fibres (Vassort 1975), cardiomyocytes (Siegelbaum and Tsien 1980; Caroni and Carafoli 1982), secretory cells (Iwatsuki and Petersen 1978), and even erythrocytes (Lew and Ferreira 1976; Grygorczyk and Schwarz (1983). The close relationship discovered between potassium and calcium channels stimulated a suggestion that the former may be a result of some modification of calcium channels under the action of penetrating Ca^{2+} ions (Hofmeier and Lux 1978). However, this suggestion has not been confirmed and it is now widely accepted that calcium-activted potassium conductance [G_K(Ca)] is related to a special type of ion channel activated from the *cytoplasmic* side of the membrane (Eckert and Tillotson 1978; Meech 1978). A good correlation between an increase in the intracellular free calcium level and activation of potassium conductance has been observed in experiments with direct measurements of $[Ca^{2+}]_{in}$ by aequorin (Eckert and Tillotson 1978). The effect of calcium ions is very specific: it cannot be reproduced by Ba^{2+} or Sr^{2+}; on the contrary, Ba^{2+} ions entering the cell depressed the potassium conductance (Gorman and Hermann 1979; Adams and Gage 1980).

Intracellular perfusion of snail neurons with calcium-buffer solution containing a fixed level of ionized calcium has shown that elevation of $[Ca^{2+}]_{in}$ to 10^{-7} M already significantly potentiates outward potassium currents (Kostyuk *et al.* 1980*a*). The additional current appeared to be not only calcium- but also potential-dependent; some membrane depolarization was necessary for its generation. In other words, intracellular Ca^{2+} ions seem to serve as a co-factor indispensable for electrical operation of the gating mechanism of these channels. This finding has been confirmed by Woolum and Gorman (1981) and Lux and Hofmeier (1982*a*, *b*). The voltage-dependence of G_K(Ca) is also present in guinea-pig hippocampal neurons (Brown and Griffith 1983), cockroach motoneurons (Thomas 1984), and other cells.

The activation time constant of calcium-dependent potassium channels reaches several milliseconds; they do not show substantial inactivation. Presumably, these channels pass several pre-activated (non-conducting) stages during activation, until membrane depolarization finally renders them into an open (conducting) state.

The pharmacological characteristics of calcium-dependent potassium channels differ from those of other potassium channels. According to our data, in snail neurons they are relatively insensitive to tetraethylammonium, a quite effective blocker of the 'usual' potassium channels (Doroshenko *et al.* 1979); this has been confirmed by experiments in other laboratories (Aldrich *et al.* 1979). However, Hermann and Gorman (1981*b*), in *Aplysia* neurons, found a very effective block of calcium-dependent potassium currents after external application of TEA, which decreased at positive membrane potentials. Intracellular injection of TEA also depressed these currents, but in this case the blocking action was potentiated by positive membrane potential shifts. Another inhibitor of voltage-operated potassium currents, 4-aminopyridine, according to the same authors (1981*a*), did not block I_K(Ca) but even potentiated it at high concentrations. The reason for such differences in the data obtained in different laboratories is not clear; perhaps this is because the neurons came from different species; it is also possible that the Ca-dependent potassium conductance is based on several types of ion channels. In fact, studies of the selectivity of this conductance have shown its considerable variability in different cells. In snail neurons the corresponding channels are very similar in this respect to the channels of the delayed potassium current; they pass monovalent cations in the sequence $K^+ = Tl^+ > Rb^+ > NH_4^+ > Cs^+ > Li^+, Na^+$ (Gorman *et al.* 1982). Channels highly selective for K^+ ions were also found in chromaffin cells, sympathetic neurons and pituitary cells (Marty 1983). But calcium-dependent potassium channels with low selectivity (which practically do not discriminate between K^+ and Na^+ ions) were found in cultured cardiomyocytes (Colquhoun *et al.* 1981). Similar channels were also described in N1E-115 neuroblastoma cells (Yellen 1982). Protons also have to be taken into account when discussing the selectivity of calcium-dependent potassium channels. Despite their low intracellular concentration, they can produce a considerable transmembrane current due to high permeability through these channels (Doroshenko and Martynyuk 1982).

The recording of single-channel activity has confirmed the suggestion about a dual mechanism of activation of calcium-dependent potassium channels, as well as about their diversity. The activation kinetics of these channels in snail neurons, like that of calcium channels, proved to be quite complex. The unitary conductance was about 19 pS, and channel openings occurred in bursts. At depolarization membrane potential shifts, channels opened with a considerable delay (7–30 ms); elevation of $[Ca^{2+}]_{in}$ removed this delay, facilitating their transition into the open state (Lux *et al.* 1981). In N1E-115 neuroblastoma cells, channels with unitary conductance of 17 pS have been recorded; their activation was potentiated by intracelluar Ca^{2+} in micromolar concentrations. In cultured rat muscle fibres, single calcium-dependent potassium channels had a much higher conductance (about 100 pS at 0°C and 300 pS at 37°C). The open-time distribution could be described by two exponents, and the frequency as well as duration of openings increased with an increase of both $[Ca^{2+}]_{in}$ and depolarization by approximately *e* times for a 15 mV potential change (Barret *et al.* 1982). The dependence on Ca^{2+} concentration of channel transitions to short or long open states was different: the short open state required the presence of at least one calcium ion, whereas the long open state required three or four ions (Magleby and Pallotta 1983*a*, *b*). Calcium-dependent potassium channels with high conductance (currents of about 10–12 pA) activated by membrane depolarization have also been described in bovine chromaffin cells, frog sympathetic neurons, and clonal pituitary cells; their open probability increased by times for a 9 mV potential change. This probability increased during elevation of $[Ca^{2+}]_{in}$ from 0.1 to 10 μM, with a parallel decrease of channel sensitivity to membrane potential shifts.

Finally, the activation kinetics of calcium-dependent potassium channels has also been studied in extracellular conditions after incorporation of membrane fragments in lipid bilayers. Under these conditions two types of calcium- and voltage-dependent processes could be observed: short channel closings (below 1 ms), producing 'bursts', and more prolonged transitions between closed and open states (time from 1 to 100 ms). In the kinetics of the second type, the mean open time appeared to be a linear function of Ca^{2+} concentration, and the mean closed time its inverse function (Moczydlowski and Latorre 1983).

An important property of calcium-dependent potassium channels, resembling the property of calcium channels, is their high dependence on intracellular cyclic nucleotide metabolism. This has been shown using perfused snail neurons, in which the introduction of the PK-A catalytic subunit potentiated I_K(Ca) (Peyer *et al.* 1982). A simultaneous decrease of free calcium level in the perfusing solution from 10 to 1 mM eliminated the effect, while its increase potentiated it. To induce the potentiating effect it was enough to inject only 0.03 μM of protein into the cell; optimal effects occurred at concentrations of 0.4–0.5 μM. All these findings led to the conclusion that the catalytic subunit increases the affinity of some group in the channel which reacts with intracellular calcium, probably by its phosphorylation.

Similar results have also been obtained on *Aplysia* neurons. All factors that increased the intracellular cAMP level produced a parallel potentiation of the calcium-dependent potassium currents; clearly, it must be cAMP that mediates the increase of the sensitivity of the corresponding channels to intracellular Ca^{2+} (Ewald and Eckert 1983). In vesicles formed from cardiomyocyte sarcolemma, the calcium-dependent transfer of K^+ was lost after dephosphorylation of membrane proteins by endogenous phosphatase. Addition of PK-A restored such transfer (Wen *et al.* 1984). Finally, the activity of corresponding channels in neuronal membrane fragments reconstituted in a lipid bilayer could also be potentiated when they were in contact with a solution containing the catalytic subunit of PK-A + ATP (Ewald and Levitan 1984).

In considering the possible functional role of calcium-dependent potassium conductance, its various properties should be taken into account. The induced outward currents have no definite threshold and do not inactivate; therefore, they may participate in processes occurring during the cell at rest. On the other hand, membrane depolarization substantially potentiates them so that these currents may be involved in intensive cellular activity. Finally, the relaxation of these currents is rather slow compared with other transmembrane currents (because they require the removal of Ca^{2+} from the inner surface of cell membrane); therefore, they may be important in the generation of long-lasting membrane phenomena.

On the basis of these properties, one may suggest that the calcium-dependent potassium conductance participates in the maintenance of resting potential and its regulation. It is known that EGTA injection in snail neurons often induces membrane depolarization accompanied by an *increase* in membrane resistance, presumably due to a decrease in the intracellular level of ionized calcium and subsequent closure of active calcium-dependent potassium channels (Alvarez-Leefmans *et al.* 1981). In natural conditions, the level of Ca^{2+} in snail neurons during rest is higher than during activity, and the cells at rest tend to be in a hyperpolarized state.

When the neuron generates a series of action potentials, the level of $[Ca^{2+}]_{in}$ rapidly increases (see Chapter 8), and this must induce a wave of hyper-polarization and stop firing. Such a combination of calcium inward currents and subsequent calcium-dependent potassium outward currents may result in the oscillatory behaviour of cells, during which periodic discharges ('bursts') will be interrupted by silent periods. Bursting activity is common for many neurons, both in invertebrate and vertebrate animals, and may have a definite functional meaning.

The generation of membrane potential oscillations with the help of calcium-dependent potassium conductance has been studied extensively, both in experiments and in mathematical models. Experimentally, an increase of $[Ca^{2+}]_{in}$ has been measured in neurons with bursting activity during the development of the 'burst'; this increase stopped at the end of the burst and at the beginning of the hyperpolarization wave. The non-linear relation between the increase of $[Ca^{2+}]_{in}$ and the induction of membrane hyperpolarization may be due to the potential-dependence of calcium-dependent potassium channels, which open more easily during membrane depolarization (Gorman *et al.* 1981). Different modes of functioning of this oscillatory system have been compared using a mathematical model which included a system of differential equations relating the changes in membrane potential and intracellular free calcium level with those in inward calcium and outward calcium-activated potassium currents and with changes in the buffering capacity of the cytoplasm. It has been shown that, depending on the parameters of the model, three types of electrical activity of the neurone can be predicted: steady hyperpolarization to the resting potential level; slow membrane potential oscillations between −30 and −60 mV with a period of 12–17 s (which is quite close to the periodicity of natural bursting activity); and steady depolarization to the level between −40 and −30 mV, during which the cell can pass into steady rhythmic activity (Mironov 1983*a*).

Figure 9.1 presents an example of electrical events calculated by this model, depending on the changes of only one parameter—the rate constant of accumulation of Ca^{2+} entering the cell by cellular buffer systems. It has been shown by the same model that a simultaneous potentiation of both calcium and potassium conductance (which can be expected during elevation of intracellular cAMP levels and increased phosphorylation of channel proteins) would result in an increase of the amplitude of membrane potential oscillations and the duration of the depolarization waves. Such effects have actually been observed in neurons during elevation of the intracellular cAMP level (Kononenko and Mironov 1980; Levitan and Norman 1980).

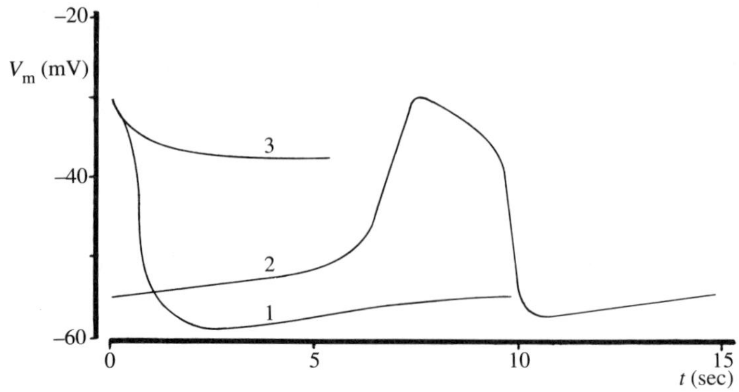

Fig. 9.1. Model presentation of electrical events on a neuronal membrane induced by the interaction of calcium and calcium-dependent potasium currents, taking into account different velocities of Ca^{2+} absorption (K) by cellular buffer systems (Mironov 1983*a*). $K = 0.1\ s^{-1}$ (1), $0.2\ s^{-1}$ (2), $0.4\ s^{-1}$ (3).

An additional factor in the induction of oscillatory behaviour can be the involvement of the system of low-threshold calcium channels. Being inactivated during weak membrane depolarization, these channels can be reactivated during hyperpolarizing potential shifts; in combination with activation of calcium-dependent potassium channels, such channels may support the generation of slow membrane potential oscillations. This mechanism can be maintained in an inactive state for a certain period (for instance, during the low resting-potential level in embryonic neurons) and then switched on at a definite time of cell development, as has been shown for developing cells of *Drosophila* larva (Wei and Salkoff 1986). The low-threshold calcium channels remain the dominant component of calcium conductance in cells that retain the capability for autorhythmic activity after maturation (cardiac pacemaker cells; Hagiwara *et al.* 1988). As already mentioned in Chapter 7, a high density of low-threshold calcium channels has been found in some hypothalamic neurons of adult rats; probably, these cells are also predisposed to spontaneous activity (Akaike *et al.* 1989). The same is characteristic for some neurons in the thalamus of rat and cat (Crunelli *et al.* 1989) and retinal bipolar neurons (Kaneko *et al.* 1989).

Finally, the calcium-dependent potassium conductance may also participate in the process of adaptation of receptor cells to long-lasting stimulation. Direct investigations of this process in stretch receptors of crayfish muscles have shown that the decrease in the amplitude of the receptor potential during continuous stretch is due to the influx of Ca^{2+} into the cell and activation of I_K(Ca) (Ottoson and Swerup 1982).

There are few data about the recurrent effect of Ca^{2+} ions on other types of membrane ion channels. In cultured mouse spinal neurons a voltage-dependent chloride conductance has been found that was also dependent on external calcium (Owen *et al.* 1984). Calcium-channel blockers inhibited this current, indicating the necessity of calcium influx into the cell for the activation of the corresponding channels. It has been suggested that this current may repolarize the cell and stabilize it at a level subthreshold for action potential generation, thus complementing the role of G_K(Ca). The calcium-dependent chloride current decayed very slowly upon membrane repolarizaton, and therefore participated in the generation of long-lasting after-hyperpolarization (Mayer 1985). A similar I_{Cl}(Ca) has also been observed in snail neurons (Walden *et al.* 1987), sensory and parasympathetic quail neurons (Bader *et al.* 1987), and lizard cone photoreceptors (Maricq and Korenbrot 1988).

Intracellular Ca^{2+} may also affect receptor-operated ion channels. Chemeris *et al.* (1982) have found that an increase of $[Ca^{2+}]_{in}$ induces inactivation of acetylcholine-operated channels in neurons of the freshwater mollusc *Planorbis*. They suggested that this effect may be responsible for the modulation of cholinergic synaptic transmission during spike activity of the cell.

Liberation of neurotransmitters and hormones

The most detailed investigations on the role of Ca^{2+} ions in the mechanism of liberation of transmitters from synaptic terminals have been carried out on frog neuromuscular junctions and squid cardiac ganglion giant synapses. The general conclusions reached from both tissues were practically identical; this allowed one to extend them to other types of synapses. More recently, another tissue became widely used—isolated synaptosomes obtained after homogenization of brain tissue; they have some advantages in the studies of calcium influx into synaptic terminals.

Several major experimental findings have laid the foundation of our understanding of the role of Ca^{2+} ions in triggering transmitter liberation.

1. The liberation is a voltage-dependent process which can, however, also take place in the absence of the sodium component in the presynaptic action potential.
2. The liberation also occurs after blocking of the voltage-operated sodium and potassium channels in the membrane of the synaptic terminal (by TTX and TEA).
3. The liberation depends critically on the presence of Ca^{2+} ions in the external medium; their specific role can be replaced only by Sr^{2+} and partly by Ba^{2+}.

4. The liberation can be considerably depressed by addition of such di- and polyvalent ions as Mg^{2+}, Mn^{2+}, Co^{2+}, and La^{3+}.

All these data led to the concept that, during depolarization of the synaptic terminal by the arriving action potential, the calcium conductance of its membrane increases due to activation of voltage-operated calcium channels. The injection of a certain amount of Ca^{2+} through these channels is the immediate reason for the exocytosis of the pre-existing synaptic vesicles filled with the transmitter.

The quantitative relations between different parts of this system have been best analysed in the squid giant synapse, in which two microelectodes could be introduced, one into the pre- and one into the postsynaptic structures, for voltage clamping and ion current measurement. The influx of Ca^{2+} in the terminal could also be measured separately by any intracellular probe.

The influx of Ca^{2+} in the terminal has been shown to decrease during strong depolarization, approaching zero at membrane potentials of about +150 mV. This 'suppression potential' corresponds to the equilibrium potential for I_{Ca} (Llinas and Nicholson 1975). The delay between the beginning of I_{Ca} and the postsynaptic response (i.e. the opening of receptor-operated channels by the liberated transmitter) was very short (about 200 μs); the relation between I_{Ca} amplitude and the amplitude of the postsynaptic response (the amount of liberated transmitter) was linear (Llinas *et al.* 1981*a*, *b*). However, in the frog neuromuscular junction this relation seemed to obey a power law (Katz and Miledi 1970).

An important question concerns the type of calcium channels operating in synaptic terminals. Initially this problem was studied on rat brain synaptosomes. Influx of ^{45}Ca was induced by high-potassium depolarization. The induction was quite effective: when 25 per cent of Na^+ ions were replaced by K^+, the ^{45}Ca influx increased by 2–3 times. The latter could be separated into a fast (1–3 s) and a slow (10–20 s) phase. Pharmacological tests indicated that these phases could be related to activation of different calcium channels. The fast phase was blocked by La^{3+} at lower concentrations, and disappeared after long treatment with high-potassium solutions; Ba^{2+} and Sr^{2+} were more effective current carriers during the fast phase. Both phases were insensitive to TTX but could be blocked by organic calcium-channel blockers (Nachshen and Blaustein 1980, 1982). However, the latter finding has not been confirmed in subsequent investigations on brain synaptosomes: the Ca^{2+} influx could not be blocked by dihydropyridines but was sensitive to ω-conotoxin; it also could not be influenced by dibutyryl-cAMP, forskolin, or phorbol esters (Freedman and Miller 1984; Suszkiw *et al.* 1986, 1989). Thus, the calcium channels in synaptic endings seemed to be different from those already studied in other neuronal membranes.

Such a comparison could be made more precisely on large synaptic terminals in which direct electrophysiological measurements are possible. In peptidergic terminals from the mammalian posterior pituitary lobe, which liberate oxytocin and vasopressin, two types of unitary calcium channels have been found (Lemos and Nowycky 1989). Channels of the first type had a high activation threshold, unitary conductance of 28 pS, and slow inactivation; they were sensitive to DHPs (L-type). Those of the second type were also high-threshold channels which inactivated rapidly and had a lower conductance (10–15 pS); they were insensitive to DHPs. The authors classified them as N_t channels. However, in frog pituitary terminals all calcium channels seem to be DHP-insensitive and can be blocked by ω-conotoxin (Obaid *et al.* 1989).

There are also unsolved questions about the intimate mechanism of coupling between the injection of Ca^{2+} in the terminal and the liberation of stored transmitter. Membrane depolarization has no direct effect on the final process; the liberation starts only when $[Ca^{2+}]_{in}$ reaches the level of several μM (Zucker and Haydon 1988). A specific role has been suggested for contractile proteins present in synaptic terminals (Puszkin *et al.* 1972). The contraction of actomyosin fibres may be related to the translocation of synaptic vesicles in the terminal and their adhesion to the inner surface of the presynaptic membrane. The synaptic terminals also contain specific proteins, called synapsins (Huttner *et al.* 1983). One of them (synapsin I) is associated with synaptic vesicles and can be effectively phosphorylated by Calm-PK (but not by PK-A). The phosphorylation of synaptic vesicles in the presence of calcium and calmodulin has also been reported by other investigators (Moskowitz *et al.* 1983). Perhaps this process is important for the transformations of synaptic vesicles that occur after injection of Ca^{2+} in the terminal.

Quite recently the idea about a possible alternative mechanism of excitation–transmitter liberation coupling has been revived, on the basis of experiments using the liberation of 'caged' Ca^{2+} ions from photosensitive Ca^{2+} chelators injected into crayfish presynaptic axons. In calcium-free bathing solution, membrane depolarization could still evoke transmitter release from such axon terminals—perhaps by inducing conformational changes in some membrane proteins that render them sensitive to Ca^{2+}; binding of Ca^{2+} to this active form then triggers the release of neurotransmitter (Hochner *et al.* 1989).

Obviously, a similar chain of processes is also characteristic for the triggering of other types of cellular secretion. This has been studied in more detail on adrenal chromaffin cells which secrete catecholamines in response to nerve impulses coming through preganglionic fibres. The presence of voltage-operated calcium channels has been shown in the membrane of these cells (Schneider *et al.* 1981; Fenwick *et al.* 1982*b*). To activate them a depolarization of the membrane is necessary, just as in synaptic terminals; this depolarization is produced by the action of the synaptic transmitter, acetylcholine. Apart from calcium channels, some calcium ions also enter the cell through the receptor-operated channels. Due to the low selectivity of these channels, they also pass divalent cations. The calcium channels in chromaffin cells are well studied and seem to be similar in the somatic membrane of different nerve cells.

Another type of well-studied secretory cells are normal and clonal malignant cells from the adenohypophysis. The presence of a calcium link in the triggering of secretion from these cells was shown many years ago (reviewed by Smith 1971; Moriarty 1978). In separate investigations of the lactotrophic (LTH) and somatotrophic cells, it was found that in the former, predominantly high-threshold calcium channels are activated and in the latter, low-threshold calcium channels are activated during depolarization (Lewis *et al.* 1988). Clonal cells of the GH_3 line secrete both hormones; the secretion can be stimulated by forskolin and phorbol esters, indicating the possible role of calcium channel phosphorylation in the triggering of secretion. Dopamine and somatostatin depressed secretion; this depression could be removed by pertussis toxin without changing the intracellular cAMP level. This may indicate the involvement of G-proteins in the regulation of calcium channel function in these cells (Boyd *et al.* 1988). However, according to the data of Johnson and Mitchell (1989) phorbol esters induced secretion of growth hormones only through activation of high-threshold calcium channels, which could be inhibited by nimodipine; the secretion of LTH seems to be induced via some other mechanism. The hypothalamic releasing hormones also modulate the activity of calcium channels in these cells. Thus, TRH induced in GH_3 cells a transitory increase of the intracellular calcium level followed by a sustained increase; the latter remained in calcium-free external solutions and so must have been due to calcium release from the intracellular stores. Membrane depolarization induced a much stronger intracellular calcium signal by activation of voltage-operated calcium channels (Benham 1989).

Hirning *et al.* (1988) have studied norepinephrine secretion from sympathetic neurons, which was related with the activation of N-type calcium channels. But in dorsal root ganglion neurons the secretion of substance P depended on the activity of DHP-sensitive L-type calcium channels (Rane *et al.* 1987).

The role of the calcium signal in triggering secretion has also been shown on pancreatic β-cells and clonal insulinoma cells (Williams 1980; Wollheim and Pozzan 1984).

Dendritic secretion

In several parts of the nervous system there is evidence of the release of substances from dendrites and, where it has been possible to study the question, this release seems to depend upon the entry of calcium ions into the dendrites. One of the best-studied systems is the substantia nigra, where the dendrites of dopaminergic neurons can release dopamine independently of action potentials in nigrostriatal axons (Chéramy *et al.* 1981) and where the release from dendrites is at least partly calcium-dependent (Geffen *et al.* 1976). A calcium-dependent release of acetylcholinesterase also occurs from dendrites of dopaminergic neurons (Greenfield *et al.* 1980; Cuello *et al.* 1981) and the release of both dopamine and acetylcholinesterase is facilitated by serotonin (Williams and Davies 1983; Taylor *et al.* 1988). The ionic basis for the release of substances from nigral dendrites probably involves the opening of a high-threshold dendritic calcium channel (Llinas *et al.* 1984; Nedergaard *et al.* 1989) which can be facilitated by application of serotonin to the dendrites but not to the cell bodies (Nedergaard *et al.* 1988). Similar calcium-dependent release of chemical substances (GABA, peptides) from dendrites probaby occurs from granule cells in the olfactory bulb (Jaffé and Cuello 1980) and from magnocellular neurons in the hypothalamus (Pow and Morris 1989).

Regulation of long-lasting processes in the nerve cells

In parallel with transient effects of Ca^{2+} ions injected into nerve cells through calcium channels, long-lasting changes in cell functioning also take place which can be classified in general terms as *cellular plasticity*. They include prolonged changes in the effectiveness of synaptic transmission, intracellular transport, the expression of genetic information, etc. The mechanisms of calcium action on these processes are much less studied, and in most cases only examples of such action can be presented.

The role of Ca^{2+} ions in the long-lasting potentiation of synaptic transmission can be illustrated by the results of model experiments carried out in our laboratory on snail neurons. As already mentioned in Chapter 6, excitatory postsynaptic responses can be evoked in some of these neurons by serotonin, via the activation of an inward current; similar effects could be induced in the same cells by intracellular injection of cAMP, indicating the possible mediation of the serotonin effect through the corresponding secondary messenger system (Konenenko *et al.* 1983, 1986). When the intracellular level of Ca^{2+} was increased in these neurons, a powerful and long-lasting potentiation of both effects occurred (Konenenko and Shcherbatko 1985*a*). The potentiation developed independently of the way through which calcium was brought into the neuron: it could be evoked by a prior series of action potentials, direct injection of Ca^{2+} through a microelectrode, or liberation of Ca^{2+} from intracellular stores (for instance, by the action of mitochondrial poisons). The potentiated inward currents remained stable for dozens of minutes after $[Ca^{2+}]_{in}$ returned to background levels (this could be checked by the disappearance of the calcium-dependent potassium currents).

Figure 9.2 demonstrates this long-lasting potentiation of serotonin- and cAMP-induced inward currents.

The biochemical mechanism of the effect is not yet known. Possibly, the synergistic action of Ca^{2+} and cAMP on the ion channels can be explained by the presence in them of two phosphorylation sites: cAMP- and Ca-calmodulin-dependent. Conformational changes induced by the second system may allosterically increase the sensitivity of the channel to cAMP. We have already met an analogous situation during the analysis of the effect of intracellular calcium on calcium-dependent potassium channels; only in this case the interaction was the opposite; cAMP acted as a conditioning factor which increased the sensitivity of the channel to the action of Ca^{2+} ions.

These effects may play an important role in plastic changes and integrative functions of the neuron by inducing long-lasting changes in the effectiveness of certain synaptic inputs after a period of increased spike activity.

Another example of plasticity is the long-lasting potentiation of synaptic transmission in some mammalian brain neurons developing during activation of the NMDA-operated channels (MacDermott *et al.* 1986; Reymann *et al.* 1989). It has been shown that the long-lasting potentiation which develops during activation of these channels depends on the elevation of $[Ca^{2+}]$ in the postsynaptic cell (Zucker 1989). Such potentiation may be most effective if the corresponding receptors and channels are located on dendritic ramifications where the rise of $[Ca^{2+}]_{in}$ is especially high and

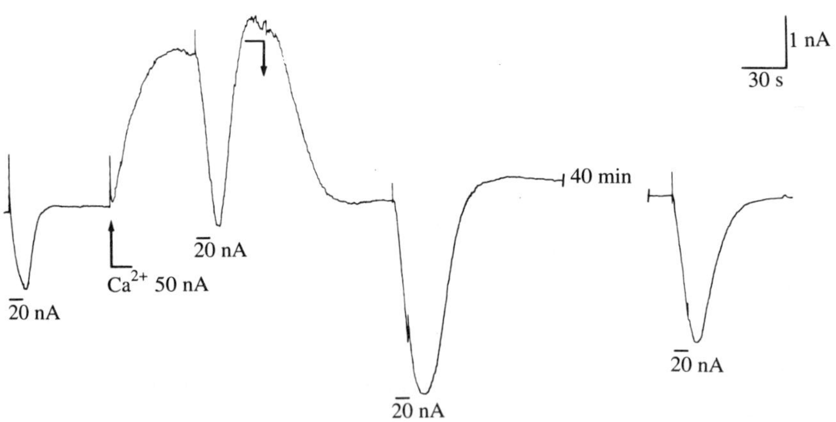

Fig. 9.2. The long-lasting effect of intracellular injection of Ca^{2+} in a snail neuron on cAMP-induced inward currents (Kononenko and Shcherbatko 1985*a*). The consecutive records were obtained before, during, and immediately after injection, and 40 min later. Amplitudes of ionophoretic currents used for injection of cAMP and Ca^{2+} are indicated near each record.

prolonged (Connor *et al.* 1988; Ross and Werman 1987). The mechanism of calcium action in this phenomenon needs further investigation.

Another important but little investigated role of Ca^{2+} ions is the regulation of intracellular transport. In nerve cells this transport is evident in the form of intra-axonal and intradendritic flow; its volume and velocity can be measured easily by distribution of radioactive substances taken by the transport system after their injection in the neuronal soma (or penetration into it through the cell membrane). The axoplasmic transport can be separated into fast and slow components. The fast transport has a velocity of about 17 mm h^{-1} (410 mm d^{-1}) in both thick and thin axons and in the axons of different vertebrates (taking into account the temperature during measurement). It does not depend on the type of substance transported (different amino acids, inorganic ions). It is generally accepted that the fast transport is related to special structural elements in the axons—microtubules and microfilaments—and depends on metabolic energy. ATP depletion reversibly locks this transport.

Although the velocity of fast transport is extremely constant, the volume of transported substances is quite variable, and Ca^{2+} ions play an important role in its regulation. This was first shown by Hammerschlag *et al.* (1975) in experiments on dorsal root ganglia isolated together with the corresponding peripheral nerve. During incubation in the solution containing a labelled amino acid (leucine), the latter is absorbed by the neurons in the ganglion and incorporated into proteins, which in turn are transported by the intra-axonal transport system. By measuring the appearance of the radioactive front in the nerve, it is possible to determine both the velocity and relative volume of transport. It was shown that after application of calcium-free solutions to the ganglion the volume of transport is decreased by 50 per cent. Changes in the concentration of Na^{+} did not affect the transport, nor did application of calcium-free solutions to the nerve fibre. Thus, Ca^{2+} ions are important for the maintenance of transport just in the neuronal soma.

A more detailed investigation of the possible mechanism of this action of calcium ions has shown that the decrease in their intracellular level does not affect the velocity of amino-acid uptake and incorporation by the cell; the main change happens at the stage of reversible binding of the synthesized substances to the transport system (Dravid and Hammerschlag 1975). Replacement of Ca^{2+} by other divalent cations has shown that Sr^{2+} can effectively substitute for Ca^{2+} but Ba^{2+} cannot. Co^{2+}, Mn^{2+}, and Mg^{2+} inhibited the action of Ca^{2+} (Hammerschlag *et al.* 1977; Ochs *et al.* 1978; Lavoie *et al.* 1979).

The intrinsic mechanism of calcium action is not quite clear. The microtubules that pass along the whole length of the axon possess small side-filaments resembling the myosin heads in muscle fibres. At the same time, the microfilaments contain actin. Possibly, the microfilaments can glide along the tubuli on the basis of an actin–myosin reaction related to ATP hydrolysis by

Ca,Mg-ATPase; the latter is, in turn, activated by the Ca-calmodulin system (Chan *et al.* 1980; Ochs 1982). Besides their participation in the axonal transporting mechanism, Ca^{2+} ions may be necessary for the 'loading' of the transported substance onto this system. These different functions of calcium can probably be separated by the action of Co^{2+} and Mg^{2+} ions. Cobalt does not affect the transport mechanism and magnesium supports its functioning; at the same time, cobalt inhibits the 'loading' whereas magnesium does not substitute for calcium in this function (Grafstein and Forman 1980).

Similar conclusions have been made by the use of another technique—the immersion of the ganglia into a solution containing labelled fucose. After uptake it is incorporated into synthesized glycoproteins (which takes place in the Golgi apparatus). The synthesis of such proteins, as well as the incorporation of fucose, was not disturbed in a solution containing 0.18 mM $CoCl_2$. However, the transport of fucosylated proteins was reversibly inhibited by about 80 per cent. If the cobalt solution was applied to a desheated part of the nerve, no changes in transport were observed distally to the place of application. This means that the cobalt-sensitive link in the transport system is located most probably at the junction between the Golgi apparatus and the microtubules (Hammerschlag and Lavoie 1979). The reversible binding of calcium to the transport system, with its subsequent movement along the axon, has also been demonstrated directly by immersion of neurons into a solution containing ^{45}Ca (Neale and Barker 1977).

Finally, the most fundamental changes in the function of a neuron under the influence of calcium signal should be expected if the latter would modify the expression of the genome, the synthesis and incorporation of channel-forming proteins and receptor molecules. Suggestions about the possibility of such mechanisms of cellular plasticity have been expressed by many authors (for example, Llinas and Sugimori 1980*a*, *b*; Lynch and Baudry 1984). The first direct evidence of the involvement of intracellular calcium in the expression of certain genes has now been published (for instance, the proenkephalin gene; Kley *et al.* 1987). On the other hand, intracellular Ca^{2+} ions definitely participate in pathological processes in nerve cells, for instance in hypoxic and ischaemic damage (Choi 1988) and in natural cell death (Ogura *et al.* 1988).

Conclusions

The functional role of calcium ions that enter the nerve cell during its activity (or are released from the intracellular stores) is extremely diverse. On the one hand, the inward calcium currents induced by membrane depolarization exert a recurrent action on membrane functions; due to slow development and powerful action on other ion channels they form a system capable of generating endogenous rhythmic activity which can, in turn, be modulated by a large number of external and internal factors. On the other hand, due to the presence in the cell of extremely effective systems for maintaining free calcium at very low levels, the injection of even small amounts of these ions during neuronal activity becomes a strong signal for the triggering of intracellular processes. Its effectiveness is multiplied through the activation of different calcium-binding proteins. By this mechanism the action of calcium can be specialized and can trigger or regulate almost all cellular functions; as a rule, the phosphorylation of key proteins by protein kinases is the most important component in these processes.

10. General conclusions

As can be seen from the findings summarized in this book, we now know quite a lot about the succession of events in which calcium participates during nerve-cell functioning. The basis of this participation—a highly asymmetric distribution of calcium ions between the extracellular medium and the cytoplasm—is a fundamental characteristic of the living state, and the influx of these ions into the cell, driven by their steep electrochemical transmembrane gradient, is an obligatory component of each cycle of neuronal activity.

Calcium fluxes through the membrane are the best-analysed events in calcium metabolism. We know the basic characteristics of different calcium channels, the density of their location in different neuronal structures, and the principles of their activation and inactivation. The mechanisms of extrusion of the excess of calcium ions back into the extracellular space are also well studied. The main point of interest now is the analysis of the molecular organization of the calcium-transporting membrane structures. So far we have only hypothetical models of the functioning of calcium channels, calcium pumps, and exchangers. Although some of the corresponding proteins have already been isolated from the cell membranes, and their amino-acid sequences have been established, we are still quite far from the understanding of how the specific functional properties of selective voltage- or receptor-operated channels are realized in these proteins. The same is true for the mechanism of calcium transport by Ca-ATPase or Na/Ca exchangers. These specific properties are formed by the secondary and tertiary structures of the corresponding protein macromolecules, and their definition is still a problem for the future.

Without knowing the secondary and tertiary structures of the channel-forming proteins, we cannot determine the nature of some important properties of calcium channels. Why is phosphorylation so important for the functioning of these channels? Where and how does the low- and high-affinity binding of cations occur in calcium channels? Why is the presence of a bound ion so important for maintaining channel properties, and why is the selectivity dramatically changed after its removal?

Without knowledge of the secondary structure of the macromolecules which form the pathways for calcium transport it is also difficult to elaborate effective pharmacological methods for the regulation of calcium metabolism and calcium-dependent cellular functions. The whole complexity of pharmacological action became obvious when it was shown that small changes in the side-groups of calcium-channel blockers may transform them into channel agonists. The number of putative binding sites for newly synthesized substances is growing all the time, and possibly soon there will be no longer space for them on the channel surface. Nevertheless, the molecular mechanisms of action of the known channel antagonists and agonists still remain unresolved.

Another question for the future concerns the role of calcium ions in the maintenance of the integrative functions of the nerve cell. There is no doubt that calcium ions are involved in practically all intracellular processes; their unique physico-chemical properties simply do not allow them to 'stay aside' from the conformational changes of complex intracellular macromolecules. Most information about the role of calcium in different metabolic processes is nowadays obtained in biochemical experiments *in vitro*; the effects observed are extremely diverse and directed to both activation and inhibition of cellular reactions. How do all these changes develop under conditions of natural cellular activity? How do they interact? The experimental conditions for the study of the role of calcium ions as 'secondary messengers' in a functioning neuron have now been created on the basis of intracellular perfusion techniques and intracellular administration of different metabolites. A wider use of these approaches will definitely help us to understand the function of neurons, and to develop effective methods for their modulation and control.

References

Adams, D. J. and Gage, P. W. (1979*a*). Ionic currents in response to membrane depolarization in an *Aplysia* neurone. *Journal of Physiology (London)* **289**, 115–41.

Adams, D. J. and Gage, P. W. (1979*b*). Characteristics of sodium and calcium conductance changes produced by membrane depolarization in an *Aplysia* neurone. *Journal of Physiology (London)* **289**, 143–61.

Adams, D. J. and Gage, P. W. (1979*c*). Sodium and calcium gating currents in an *Aplysia* neurone. *Journal of Physiology (London)* **291**, 467–81.

Adams, D. J. and Gage, P. W. (1980). Divalent ion currents and the delayed potassium conductance in an *Aplysia* neurone. *Journal of Physiology (London)* **304**, 297–313.

Ahmed, Z. and Connor, J. A. (1979). Measurements of calcium influx under voltage clamp in molluscan neurones using the metallochromic dye arsenazo III. *Journal of Physiology (London)* **286**, 61–82.

Ahmed, Z. and Connor, J. A. (1988). Calcium regulation by and buffer capacity of molluscan neurons during calcium transients. *Cell Calcium* **9**, 57–69.

Akaike, N., Lee, K. S., and Brown, A. M. (1978). The calcium current of *Helix* neuron. *Journal of General Physiology* **71**, 509–31.

Akaike, N., Nishi, K., and Oyama, Y. (1983). Characteristics of manganese current and its comparison with current carried by other divalent cations in snail soma membrane. *Journal of Membrane Biology* **76**, 289–97.

Akaike, N., Tsuda, Y., and Oyama, Y. (1988). Separation of current- and voltage-dependent inactivation of calcium current in frog sensory neurons. *Neuroscience Letters* **84**, 46–50.

Akaike, N., Kostyuk, P. G., and Osipchuk, Y. V. (1989). Dihydropyridine-sensitive low-threshold calcium channels in isolated rat hypothalamic neurones *Journal of Physiology (London)* **412**, 181–95.

Akopyan, A. R., Chemeris, N. K., and Iljin, V. I. (1985). Neurotransmitter-induced modulation of neuronal Ca current is not mediated by intracellular Ca or cAMP. *Brain Research* **326**, 145–8.

Albus, U., Habermann, E., Ferry, D. R., and Glossmann, H. (1984). Novel 1,4-dihydropyridine (Bay K 8644) facilitates calcium-dependent (^{3}H)-noradrenaline release from PC 12 cells. *Journal of Neurochemistry* **42**, 1186–9.

Aldrich, R. W., Getting, P. A., and Thompson, S. H. (1979). Inactivation of delayed outward current in molluscan neuron somata. *Journal of Physiology (London)* **291**, 507–30.

Allen, C. N., Brady, R., Swan, J., Hori, N., and Carpenter, D. O. (1988). *N*-methyl-D-aspartate (NMDA) receptors are inactivated by trypsin. *Brain Research* **458**, 147–50.

Almers, W. and McCleskey, E. W. (1984). Non-selective conductance in calcium channels of frog muscle: calcium selectivity in a single-file pore. *Journal of Physiology (London)* **353**, 585–608.

Almers, W., McCleskey, E. W., and Palade, P. T. (1984). A non-selective cation conductance in frog muscle membrane blocked by micromolar external calcium ions. *Journal of Physiology (London)* **353**, 565–83.

Alvarez, O., Brodwick, M., Latorre, R., McLaughlin, A., McLaughlin, S., and Szabo, G. (1983). Large divalent cations and electrostatic potentials adjacent to membranes: Experimental results with hexamethonium. *Biophysical Journal* **44**, 333–42.

Alvarez-Leefmans, F. J. and Miledi, R. (1980). Voltage sensitive calcium entry in frog motoneurones. *Journal of Physiology (London)* **303**, 241–57.

Alvarez-Leefmans, F. J., Rink, T. J., and Tsien, R. Y. (1981). Free calcium ions in neurons of *Helix aspersa* measured with ion-selective microelectrodes. *Journal of Physiology (London)* **315**, 531–48.

Anderson, A. J. and Harvey, A. L. (1987). ω-Conotoxin does not block the verapamil-sensitive calcium channels at mouse motor nerve terminals. *Neuroscience Letters* **82**, 177–80.

Anderson, C. L., Scriabine, A., and Janis, R. A. (1984). Effects of a Ca^{2+} channel agonist (Bay K 8644) and antagonists (nitrendipine, nisoldipine, verapamil and diltiazem) in Ca^{2+} uptake by rabbit aortic rings. *Federation Proceedings* **43**, 550.

Anderson, K., Lai, F. A., Liu, Q.-Y., Rosseau, E., Erickson, H. P., and Meissner, G. (1989). Structural and functional characterization of the purified cardiac ryanodine receptor-Ca^{++} release channel complex. *Journal of Biological Chemistry* **264**, 1329–35.

Anderson, N. S., Ramon, F., and Snyder, A. (1971). Studies on calcium and sodium currents in uterine smooth muscle excitation under current-clamp and voltage-clamp conditions. *Journal of General Physiology* **58**, 322–39.

Armstrong, C. M. and Bezanilla, F. (1973). Currents related to the movement of the gating particles of the sodium channel. *Nature* **242**, 459–61.

Armstrong, C. M., Bezanilla, F. M., and Horowicz, R. (1972). Twitches in the presence of ethylene glycol-*bis*(β-aminoethyl ether)-*N*,*N*′-tetraacetic acid. *Biochimica et Biophysica Acta* **267**, 605–8.

Armstrong, C. M., Bezanilla, F., and Rojas, E. (1973). Destruction of sodium conductance inactivation in squid axons perfused with pronase. *Journal of General Physiology* **62**, 375–91.

Armstrong, D. L. (1989). Calcium channel regulation by calcineurin, a Ca^{++} activated phosphatase in mammalian brain. *Trends in Neurosciences* **12**, 1–10.

Armstrong, D. and Eckert, R. (1985). Phosphorylating agents prevent washout of unitary calcium currents in excised membrane patches. *Journal of General Physiology* **86**, 25a.

Armstrong, D. and Eckert, R. (1987). Voltage-activated calcium channels that must be phosphorylated to respond to membrane depolarization. *Proceedings of the National Academy of Sciences USA* **84**, 2518–22.

Armstrong, D. and Kalman, D. (1988). The role of protein phosphorylation in the response of dihydropyridine-sensitive calcium channels to membrane depolarization in mammalian pituitary tumor cells. In *Calcium and ion channel modulation*, Proceedings of a Symposium honouring R. Eckert, 26 February-1 March, 1987, Los Angeles, California (ed. A. D. Grinnel, D. Armstrong, and M. B. Jackson), pp. 215–27. Plenum, New York.

Aschcroft, F. M. and Stanfield, P. R. (1982). Calcium inactivation in skeletal muscle fibres of the stick insect, *Carausius morosus*.*Journal of Physiology (London)* **330**, 349–72.

Ascher, P. and Nowak, L. (1987). Electrophysiological studies of NMDA receptors. *Trends in Neurosciences* **10**, 284–7.

Ascher, P. and Nowak, L. (1988). The role of divalent cations in the *N*-methyl-D-aspartate responses of mouse central neurones in culture. *Journal of Physiology (London)* **399**, 247–66.

Bader, C. R., Bertrand, D., and Schlichter, R. (1987). Calcium-activated choride current in cultured sensory and parasympathetic quail neurones. *Journal of Physiology (London)* **394**, 125–48.

Baker, P. F. (1976). The regulation of intracellular calcium. In *Calcium in Biological Systems*, Symposium of the Society for Experimental Biology, No. 30, (ed. C. J. Duncan) pp. 67–88. Cambridge University Press.

Baker, P. F. and McNaughton, P. A. (1976). Kinetics and energetics of calcium efflux from intact squid giant axons. *Journal of Physiology (London)* **259**103–44.

Baker, P. F. and Umbach, J. A. (1983). Calcium electrode determination of ionized Ca and Ca-buffering capacity of squid axoplasm. *Journal of Physiology (London)* **341**, 61P.

Baker, P. F., Hodgkin, A. L., and Ridgway, E. B. (1971). Depolarization and calcium entry in squid giant axons. *Journal of Physiology (London)* **218**, 709–55.

Baker, P. F., Meves, H., and Ridgway, E. B. (1973*a*). Effects of manganese and other agents on the calcium uptake that follows depolarization of squid axons. *Journal of Physiology (London)* **231**, 511–26.

Baker, P. F., Meves, H., and Ridgway, E. B. (1973*b*). Calcium entry in response to maintained depolarization of squid axons. *Journal of Physiology (London)* **231**, 527–48.

Barhanin, J., Coppola, T., Schmid, A., Borsotto, M., and Lazdunski, M. (1987). The calcium channel antagonists receptor from brain skeletal muscle. Reconstruction after purification and subunit characterization. *European Journal of Biochemistry* **164**, 525–31.

Barhanin, J., Schmid, A., and Lazdunski, M. (1988). Properties of structure and interaction of the receptor for ω-conotoxin, a polypeptide active on Ca^{2+} channels. *Biochemical and Biophysical Research Communications* **150**, 1051–62.

Barret, E. and Barret, J. (1976). Separation of two voltage-sensitive potassium currents and demonstration of a tetrodotoxin-resistant calcium current in frog motoneurones. *Journal of Physiology (London)* **255**, 737–74.

Barret, J. N., Magleby, K. L., and Pallotta, B. (1982). Properties of single calcium-activated potassium channels in cultured rat muscle. *Journal of Physiology (London)* **331**, 211–30.

Beam, K. G. and Knudson, C. M. (1988). Calcium currents in embryonic and neonatal mammalian skeletal muscle. *Journal of General Physiology* **91**, 781–98.

Bean, B. P. (1985). Two kinds of calcium channels in canine atrial cells. Differences in kinetics, selectivity and pharmacology. *Journal of General Physiology* **86**, 1–30.

Begenisich, T. (1975). Magnitude and location of surface charges on *Myxicola* giant axons. *Journal of General Physiology* **66**, 47–65.

Belan, P. V., Osipenko, O. N., and Tepkin, A. V. (1989). Calcium release from intracellular stores of the neuronal soma of the mollusc inducted by inositol triphosphate and nonhydrolyzed GTP analogue. *Neurophysiology (Kiev)* **21**, 707–10.

Belardetti, F., Schacher, S., and Siegelbaum, S. A. (1986). Action potentials, macroscopic and single channel currents recorded from growth cones of *Aplysia* neurones in culture. *Journal of Physiology (London)* **374**, 289–313.

Bellemann, P., Ferry, D., Luebbecke, F., and Glossmann, H. (1981). (^{3}H)-nitrendipine, a potent calcium antagonist, binds with high affinity to cardiac membranes. *Arzneimittel-Forschung* **31**, 2064–7.

Bellemann, P., Schade, A., and Towart, R. (1983). Dihydropyridine receptor in rat brain labeled with (^{3}H)-nimodipine. *Proceedings of the National Academy of Science USA* **80**, 2356–60.

Belles, B., Malecot, C. O., Hescheler, J., and Trautwein, W. (1988). "Run-down" of the Ca current during long whole-cell recordings in guinea pig heart cells: role of phosphorylation and intracellular calcium. *Pfluegers Archiv* **411**, 353–60.

Benham, C. D. (1989). Voltage-gated and agonist-mediated rises in intracellular Ca^{++} in rat clonal pituitary cells (GH^3) held under voltage clamp. *Journal of Physiology (London)* **415**, 143–58.

Benham, C. D. and Tsien, R. W. (1987). A novel receptor-operated Ca^{2+}-permeable channel activated by ATP in smooth muscle. *Nature* **328**, 275–8.

Benham, C. D. and Tsien, R. W. (1988). Noradrenaline modulation of calcium channels in single smooth muscle cells from rabbit ear artery. *Journal of Physiology (London)* **404**, 767–84.

Bernal, M. J., Fernandez, J. R., and Alvarez-Leefmans, F. J. (1987). Trifluoperazine blocks Ca^{2+} inward currents in *Helix* neurons in a reversible and dose dependent manner. *Biophysical Journal* **51**, (2), 33.

Bernstein, J. (1912). *Electrobiologie* Braunswchweig: Vieweg.

Berridge, M. J. (1988). Inositol trisphosphate-induced membrane potential oscillations in *Xenopus* oocytes. *Journal of Physiology (London)* **403**, 589–99.

Blaustein, M. P. (1976). The ins and outs of calcium transport in squid axons: internal and external ion activation of calcium efflux. *Federation Proceedings* **35**, 2574–8.

Blaustein, M. P. (1977). Effects of internal and external cations and ATP on sodium–calcium and calcium–calcium exchange in squid axons. *Biophysical Journal* **20**, 79–111.

Blaustein, M. P. (1988). Calcium transport and buffering in neurons. *Trends in Neurosciences* **11**, 438–43.

Blaustein, M. P. and Goldman, D. E. (1968). The action of certain polyvalent cations in the voltage-clamped lobster axon. *Journal of General Physiology* **51**, 279–97.

Blaustein, M. P. and Hodgkin, A. L. (1969). The effect of cyanide on the efflux of calcium from squid axons. *Journal of Physiology (London)* **200**, 497–527.

Blaustein, M. P., Ratzlaff, R. W., Kenarick, N. C., and Schwetzer, E. S. (1978*a*). Calcium buffering in presynaptic nerve terminals. I. Evidence for involvement of a nonmitochondrial ATP-dependent sequestration mechanism. *Journal of General Physiology* **72**, 15–41.

Blaustein, M. P., Ratzlaff, R. W., and Schwetzer, E. S. (1978*b*). Calcium buffering in presynaptic nerve terminals. II. Kinetic properties of the nonmitochondrial Ca-sequestration mechanism. *Journal of General Physiology* **72**, 43–66.

Blaustein, M. P., Ratzlaff, R. W., and Schwetzer, E. S. (1980). Control of intracellular calcium in presynaptic nerve terminals. *Federation Proceedings* **39**, 2790–5.

Bodewei, R., Hering, S., Schubert, B., and Wollenberger, A. (1985). Sodium and calcium currents in neuroblastoma × glioma hybrid cells before and after morphological differentiation by dibutyryl cyclic AMP. *General Physiology and Biophysics* **4**, 113–27.

Boll, W. and Lux, H. D. (1985). Action of organic antagonists on neuronal calcium currents. *Neuroscience Letters* **56**, 335–9.

Bolognani, L. *et al.* (1981). Lipid composition in ganglia of molluscs. *Journal of Neurochemistry* **36**, 821–5.

Bonvallet, R. (1987). A low threshold calcium current recorded at physiological Ca concentrations in single frog atrial cells. *Pfluegers Archiv* **408**, 540–2.

Borsotto, M., Barhanin, J., Fosset, M., and Lazdunski, M. (1985). The 1,4-dihydropyridine receptor associated with the skeletal muscle voltage-dependent Ca^{2+} channel. *Journal of Biological Chemistry* **260**, 14255–63.

Bossu, J.-L. and Feltz, A. (1986). Inactivation of the low-threshold transient calcium current in rat sensory neurones: evidence for a dual process. *Journal of Physiology (London)* **376**, 341–57.

Bossu, J.-L., Feltz, A., and Thomann, J. M. (1985). Depolarization elicits two distinct calcium currents in vertebrate sensory neurones. *Pfluegers Archiv* **403**, 360–8.

Bossu, J.-L., Dupont, J.-L., and Feltz, A. (1989*a*). Calcium currents in rat cerebeller Purkinje cells maintained in culture. *Neuroscience* **30**, 605–17.

Bossu, J.-L., Fagni, L., and Feltz, A. (1989*b*). Voltage-activated calcium channels in rat Purkinje cells maintained in culture. *Pfluegers Archiv* **414**, 92–4.

Bossu, J.-L., Rodeau, J. L., and Feltz, A. (1989*c*). Decay kinetics of calcium currents in rat sensory neurones: analysis at two internal free calcium concentrations. *Pfluegers Archiv* **414**, 89–91.

Boyd, R. S., Ray, K. P., and Wallis, M. (1988). Actions of pertussis toxin on the inhibitory effects of dopamine and somatostatin on prolactin and growth hormone release from ovine anterior pituitary cells. *Journal of Molecular Endocrinology* **1**, 179–86.

Bozem, M., Henquin, M., and Henquin, J.-C. (1987). The ionic, electrical, and secretory effects of protein kinase C activation in mouse pancreatic B-cells: studies with a phorbol ester. *Endocrinology* **121**, 1025–33.

Brehm, P., Eckert, R., and Tillotson, D. (1980). Calcium-mediated inactivation of calcium current in *Paramecium*. *Journal of Physiology (London)* **306**, 193–203.

Brezina, V., Eckert, R., and Erxleben, C. (1987). Suppression of calcium current by an endogenous neuropeptide in neurones of *Aplysia californica*. *Journal of Physiology (London)* **388**, 565–95.

Brinley, F. J., Tiffert, T., Scarpa, A., and Mullins, L. J. (1977). Intracellular calcium buffering capacity in isolated squid axons. *Journal of General Physiology* **70**, 355–84.

Brodwick, M. S. and Eaton, D. C. (1978). Sodium channel inactivation in squid axon is removed by high internal pH or trypsin-specific reagents. *Science* **200**, 1494–6.

Brown, A. M., Morimoto, K., Tsuda, Y., and Wilson, D. L. (1981). Calcium current-dependent and voltage-dependent inactivation of calcium channels in *Helix aspersa*. *Journal of Physiology (London)* **320**, 193–218.

Brown, A. M., Tsuda, Y., and Wilson, D. L. (1983). A description of activation and conduction in calcium channels based on tail and turn-on current measurements in the snail. *Journal of Physiology (London)* **344**, 549–83.

Brown, A. M., Lux, H. D., and Wilson, D. L. (1984). Activation and inactivation of single calcium channels in snail neurones. *Journal of General Physiology* **83**, 751–69.

Brown, A. M., Kunze, D. L., and Yatani, A. (1986). Dual effects of dihydropyridines on whole cell and unitary calcium currents in single ventricular cells of guinea-pig. *Journal of Physiology (London)* **379**, 495–514.

Brown, D. A. and Griffith, W. H. (1983). Calcium-activated outward current in voltage-clamped hippocampal neurones of the guinea-pig. *Journal of Physiology (London)* **337**, 287–301.

Bülbring, E. and Tomita, T. (1969). Effect of calcium, barium and manganese in the smooth muscle of the guinea-pig taenia coli. *Proceedings of the Royal Society of London* **B172**, 121.

Burgess, A. J. and Norman, R. I. (1988). Characterization of calcium channel subunit. *Biochemical Society Transactions* **16**, 452–4.

Bury, V. A. and Shuba, M. F. (1976). Transmembrane ionic currents in smooth muscle cells of ureter during excitation. In *Physiology of smooth muscle* (ed. E. Bulbring and M. F. Shuba), pp. 65–75. Raven Press, New York.

Byerly, L. and Hagiwara, S. (1982). Calcium currents in internally perfused nerve cell bodies of *Lymnea stagnalis*. *Journal of Physiology (London)* **322**, 503–28.

Byerly, L. and Moody, W. J. (1984). Intracellular calcium ions and calcium currents in perfused neurons of the snail, *Lymnaea stagnalis*. *Journal of Physiology (London)* **352**, 637–52.

Byerly, L. and Yazejian, B. (1986). Intracellular factors for the maintenance of calcium currents in perfused neurones from the snail, *Lymnaea stagnalis*. *Journal of Physiology (London)* **370**, 631–50.

Byerly, L., Chase, P. B., and Stimers, J. R. (1984). Calcium current activation kinetics in neurones of the snail *Lymnaea stagnalis*. *Journal of Physiology (London)* **348**, 187–207.

Byerly, L., Leung, H.-T., and Yazejian, B. (1988). Cellular control of calcium currents. *Biomedical Research* **9**, (2), 1–9.

Callewaert, G., Hanbauer, I., and Morad, M. (1989). Modulation of calcium channels in cardiac and neuronal cells by an endogenous peptide. *Science* **243**, 663–6.

Campbell, K. P., Leung, A. T., and Sharp, A. H. (1988). The biochemistry and molecular biology of the dihydropyridine-sensitive calcium channel. *Trends in Neurosciences* **11**, 425–30.

Carafoli, E. (1988). Intracellular calcium regulation, with special attention to the role of the plasma membrane calcium pump. *Journal of Cardiovascular Pharmacology* **12**, (3), S77–S84.

Carafoli, E. and Zurini, M. (1982). The Ca^{2+}-pumping ATPase of plasma membranes: purification, reconstitution and properties. *Biochimica et Biophysica Acta* **683**, 279–301.

Carbone, E. and Lux, H. D. (1984). A low voltage-activated, fully inactivating Ca channel in vertebrate sensory neurones. *Nature* **310**, 501–3.

Carbone, E. and Lux, H. D. (1987*a*). Kinetics and selectivity of a low-voltage-activated calcium current in chick and rat sensory neurones. *Journal of Physiology (London)* **386**, 547–70.

Carbone, E. and Lux, H. D. (1987*b*). Single low-voltage-activated calcium channels in chick and rat sensory neurones. *Journal of Physiology (London)* **386**, 571–601.

Carbone, E. and Lux, H. D. (1988). ω-Conotoxin blockade distinguishes Ca from Na permeable states in neuronal calcium channels. *Pfluegers Archiv* **413**, 14–22.

Caroni, P. and Carafoli, E. (1982). Modulation by calcium of the potassium permeability of dog heart sarcolemmal vesicles. *Proceedings of the National Academy of Sciences USA* **79**, 5763–7.

Caroni, P. and Carafoli, E. (1983). The regulation of the Na^+–Ca^{2+} exchanger of heart sarcolemma. *European Journal of Biochemistry* **132**, 451–60.

Caroni, P., Reinlib, L., and Carafoli, E. (1980). Charge movements during the Na^+–Ca^{2+} exchange in heart sarcolemmal vesicles. *Proceedings of the National Academy of Sciences USA* **77**, 6354–8.

Catterall, W. A. (1989). Molecular properties of voltage-sensitive sodium and calcium channels. *Journal of Cellular Biochemistry* Suppl. 13 E, p. 168.

Cauvin, C., Loutzenhiser, R., and van Breemen, C. (1983). Mechanism of calcium antagonist-induced vasodilation. *Annual Review of Pharmacology and Toxicology* **23**, 373–96.

Cavalie, A., Ochi, R., Pelzer, D., and Trautwein, W. (1983). Elementary currents through Ca^{2+} channels in guinea pig myocytes. *Pfluegers Archiv* **398**, 284–97.

Cavalie, A., Pelzer, D., and Trautwein, W. (1986). Fast and slow gating behaviour of single calcium channels in cardiac cells. Relation to activation and inactivation of calcium-channel current. *Pfluegers Archiv* **406**, 241–58.

Cavalie, A., Flockerzi, V., Hofmann, F., Pelzer, D., and Trautwein, W. (1987). Two types of calcium channels from rabbit fast skeletal muscle transverse tubules in liquid bilayers: differences in conductance, gating behaviour and chemical modulation. *Journal of Physiology (London)* **390**, 82P.

Chad, J. E. and Eckert, R. (1984*a*). Calcium domains associated with individual channels can account for anomalous voltage relations of Ca-dependent responses. *Biophysical Journal* **45**, 993–9.

Chad, J. E. and Eckert, R. (1984*b*). Stimulation of cAMP-dependent protein phosphorylation retards both inactivation and 'washout' of Ca current in dialyzed *Helix* neurons. *Society for Neuroscience Abstracts* **10**, 866.

Chad, J. E. and Eckert, R. (1986). An enzymatic mechanism for calcium current inactivation in dialysed *Helix* neurones. *Journal of Physiology (London)* **378**, 31–51.

Chagovetz, V. Yu. (1903–1906). Essay on electric events in living tissues from the viewpoint of newest physico-chemical theories. *StPt* **1–2**, 475 (in Russian).

Chan, S. Y., Ochs, S., and Worth, R. M. (1980). The requirement for calcium ions and the effect of other ions on axoplasmic transport in mammalian nerve. *Journal of Physiology (London)* **301**, 477–504.

Chapman, D. L. (1913). A contribution to the theory of electrocapillarity. *Philosophical Magazine* **25**, 475–81.

Chemeris, N. K., Kazachenko, V. N., Kislov, A. N., and Kurchikov, A. L. (1982). Inhibition of acetylcholine responses by intracellular calcium in *Lymnaea stagnalis* neurones. *Journal of Physiology (London)* **323**, 1–19.

Chen, C., Corbley, M. J., Roberts, T. M., and Hess, P. (1988). Voltage-sensitive calcium channels in normal and transformed 3T3 fibroblasts. *Science* **239**, 1024–6.

Chéramy, A., Leviel, V., and Glowinski, J. (1981). Dendritic release of dopamine in the substantia nigra. *Nature* **289**, 537–42.

Chiesi, M. and Carafoli, E. (1982). The regulation of Ca^{2+} transport by fast skeletal muscle sarcoplasmic reticulum: Role of calmodulin and of the 53,000-dalton glycoprotein. *Journal of Biological Chemistry* **257**, 984–91.

Choi, D. W. (1988). Calcium-mediated neurotoxicity: relationship to specific channel types and role in ischemic damage. *Trends in Neurosciences* **11**, 465–9.

Clapham, D. E. and Neher, E. (1984). Trifluoperasine reduces inward ionic currents and secretion by separate mechanisms in bovine chromaffin cells. *Journal of Physiology (London)* **353**, 541–64.

Clark, M. G., Gannon, B. J., Bodkin, N., Patten, G. S., and Berry, M. N. (1978). An improved procedure for the high-yield preparation of intact heart cells from the adult rat: Biochemical and morphological study. *Journal of Molecular and Cellular Cardiology* **10**, 1101–21.

Cognard, C., Lazdunski, M., and Romey, G. (1986). Different types of Ca^{2+} channels in mammalian skeletal muscle cells in culture. *Proceedings of the National Academy of Sciences USA* **83**, 517–21.

Cohen, C. J. and Chung, M. (1984). A nifedipine-derivative (Bay K 8644) that increases Ca-currents in myocardial cells: A novel positive ionotropic agent. *Biophysical Journal* **45**, 394A.

Cohen, C. J. and McCarthy, R. T. (1987). Nimodipine block of calcium channels in rat anterior pituitary cells. *Journal of Physiology (London)* **387**, 195–225.

Cohen, C. J., McCarthy, R. T., Barrett, P. Q., and Rasmussen, H. (1988). Ca channels in adrenal glomerulosa cells: K^+ and angiotensin II increase T-type Ca channel current. *Proceedings of the National Academy of Sciences USA* **85**, 2412–16.

Colquhoun, D. and Sakmann, B. (1981). Fluctuations in the microsecond time range of the current through single acetylcholine receptor ion channels. *Nature* **294**, 464–6.

Colquhoun, D., Neher, E., Reuter, H., and Stevens, C. F. (1981). Inward current channels activated by intracellular Ca in cultured cardiac cells. *Nature* **294**, 752–4.

Connor, J. A. (1977). Time course separation of two inward currents in molluscan neurons. *Brain Research* **119**, 487–92.

Connor, J. A. (1979). Calcium current in molluscan neurones: Measurement under conditions which maximize its visibility. *Journal of Physiology (London)* **286**, 41–60.

Connor, J. A., Wadman, W. J., Hockberger, P. E., and Wong, R. K. S. (1988). Sustained dendritic gradients of Ca^{2+} induced by excitatory amino acids in CA1 hippocampal neurons. *Science* **240**, 649–53.

Coraboeuf, E. and Otsuka, M. (1956). L'action des solutions hyposodiques sur les potentiels cellulaires de tissu cardiaque de mammifères. *Comptes Rendus de l'Academie des Sciences* **243**, 441–4.

Coray, A., *et al.* (1980). Resting calcium in sheep cardiac tissue and in frog skeletal muscle measured with ion-selective micro-electrode. *Journal of Physiology (London)* **305**, 60P–61P.

Cota, G. (1986). Calcium channel currents in pars intermedia cells of the rat pituitary gland. Kinetic properties and washout during intracellular dialysis. *Journal of General Physiology* **88**, 83–105.

Cota, G. and Stefani, E. (1984). Saturation of calcium channels and surface charge effects in skeletal muscle fibres of the frog. *Journal of Physiology (London)* **351**, 135–54.

Cota, G. and Stefani, E. (1986). A fast-activated inward calcium current in twitch muscle fibres of the frog (*Rana montezume*). *Journal of Physiology (London)* **370**, 151–63.

Crunelli, V. and Pollard, C. E. (1988). A transient inward calcium current in cat and rat thalamic cells *in vitro*. *Journal of Physiology (London)* **407**, 45P.

Crunelli, V., Lightowler, S., and Pollard, C. E. (1989). A T-type Ca^{2+} current underlies low-threshold Ca^{2+} potentials in cells of the cat and rat lateral geniculate nucleus. *Journal of Physiology (London)* **413**, 543–61.

Cruz, L. J., Johnson, D. S., and Olivera, B. M. (1987). Characterization of the ω-conotoxin target. Evidence for tissue-specific heterogeneity in calcium channel types. *Biochemistry* **26**, 820–4.

Cuello, A. C., Romero, E., and Smith, A. D. (1981). *In vitro* release of acetylcholinesterase from the rat substantia nigra. *Journal of Physiology (London)* **312**, 14P–15P.

Curtis, B. A. and Prosser, C. L. (1977). Calcium and cat intestinal smooth muscle. In *Excitation–contraction coupling in smooth muscle* (ed. R. Casteels *et al.*), pp. 123–9. Elsevier, Amsterdam.

Curtis, B. M. and Catterall, W. A. (1984). Purification of the calcium antagonist receptor of the voltage-sensitive calcium channel from skeletal muscle transverse tubules. *Biochemistry* **23**, 2113–18.

Davies, N. W., Lux, H. D., and Morad, M. (1988). Site and mechanism of activation of proton-induced sodium current in chick dorsal root ganglion neurones. *Journal of Physiology (London)* **400**, 159–87.

Deisz, R. A. and Lux, H. D. (1985). γ-Aminobutyric acid-induced depression of calcium currents of chick sensory neurons. *Neuroscience Letters* **56**, 205–10.

Deitmer, J. W. (1986). Voltage dependence of two inward currents carried by calcium and barium in the ciliate *Stylonychia mytilus*. *Journal of Physiology (London)* **380**, 551–74.

Deitmer, J. W., Ivens, I., and Pernberg, J. (1986). Changes in voltage-dependent calcium currents during the cell cycle of the ciliate *Stylonychia*. *Experimental Cell Research* **162**, 549–54.

DeRiemer, S. A., Strong, J. A., Albert, K. A., Greengard, P., and Kaczmarek, L. K. (1985). Enhancement of calcium current in *Aplysia* neurones by phorbol ester and protein kinase C. *Nature* **313**, 313–16.

Dichter, M. A. and Zona, C. (1989). Calcium currents in cultured rat cortical neurons. *Brain Research* **492**, 219–29.

Dichter, M. A., Lisak, J., and Biales, R. (1983). Action potential mechanism of mammalian cortical neurons in cell culture. *Brain Research* **289**, 99–107.

DiPolo, R. (1973). Calcium efflux from internally dialyzed squid giant axons. *Journal of General Physiology* **62**, 575–89.

DiPolo, R. (1977). Characterization of the ATP-dependent calcium efflux in dialyzed squid giant axons. *Journal of General Physiology* **69**, 795–813.

DiPolo, R. and Beaugé, L. (1983). The calcium pump and sodium–calcium exchange in squid axons. *Annual Review of Physiology* **45**, 313–24.

DiPolo, R. *et al.* (1976). Ionized calcium concentrations in squid axons. *Journal of General Physiology* **67**, 433–67.

DiPolo, R., Caputo, C., and Bezanilla, F. (1983). Voltage-dependent calcium channel in the squid axon (late Ca channel). *Proceedings of the National Academy of Sciences USA* **80**, 1743–5.

Doerner, D. and Alger, B. E. (1988). Cyclic GMP depresses hippocampal Ca^{2+} current through a mechanism independent of cGMP-dependent protein kinase. *Neuron* **1**, 693–9.

Dolgaya, E. V. and Mironov, S. L. (1984). Properties of the neuronal surface of rat dorsal root ganglia studied by microelectrophoresis. *Neurophysiology (Kiev)* **16**, 176–82.

Dolphin, A. C. and Scott, R. H. (1987). Inhibition of calcium currents by the GABA agonist baclofen: involvement of a nucleotide binding protein. *Neuroscience Letters* Suppl. **29**, 85.

Dolphin, A. C. and Scott, R. H. (1989). Interaction between calcium channel ligands and guanine nucleotides in cultured rat sensory and sympathetic neurones. *Journal of Physiology (London)* **413**, 271–88.

Dolphin, A. C., Forda, S. R., and Scott, R. H. (1986). Calcium-dependent currents in cultured rat dorsal root ganglion neurones are inhibited by an adenosine analogue. *Journal of Physiology (London)* **373**, 47–61.

Donaldson, P. L. and Beam, K. G. (1983). Calcium currents in a fast-twitch skeletal muscle of the rat. *Journal of General Physiology* **67**, 433–67.

Doroshenko, P. A. and Kostyuk, P. G. (1987). Enhancement of calcium current in the somatic membrane of snail nerve cells by phorbol ester. *Biological Membranes (Moscow)* **4**, 1160–3.

Doroshenko, P. A. and Martynyuk, A. E. (1982). A blockade of Ca^{2+}-dependent outward current in the somatic membrane of molluscan neurons by increasing intracellular pH. *Neurophysiology (Kiev)* **14**, 426–30.

Doroshenko, P. A. and Martynyuk, A. E. (1984). The effect of calmodulin blockers on inhibition of potential-dependent calcium conductance in the nerve cell by intracellular Ca^{2+} ions. *Doklady Akademii Nauk SSSR (Moscow)* **274**, 471–3.

Doroshenko, P. A. and Tsyndrenko, A. Ya. (1978). The action of intracellular calcium on calcium inward current. *Neurophysiology (Kiev)* **10**, 203–5.

Doroshenko, P. A., Kostyuk, P. G., and Tsyndrenko, A. Ya. (1978). Separation of potassium and calcium channels in the somatic membrane of nerve cell. *Neurophysiology (Kiev)* **10**, 645–53.

Doroshenko, P. A., Kostyuk, P. G., and Tsyndrenko, A. Ya. (1979). The study of TEA-resistant outward current in the somatic membrane of perfused nerve cells. *Neurophysiology (Kiev)* **11**, 460–8.

Doroshenko, P. A., Kostyuk, P. G., and Martynyuk, A. E. (1982). Intracellular metabolism of aenosine 3′,5′-cyclic monophosphate and calcium inward current in perfused neurones of *Helix pomatia*. *Neuroscience* **7**, 2125–34.

Doroshenko, P. A., Kostyuk, P. G., and Martynyuk, A. E. (1984*a*). Inactivation of calcium current in the somatic membrane of snail neurons. *General Physiology and Biophysics* **3**, 1–17.

Doroshenko, P. A., Kostyuk, P. G., Martynyuk, A. E., Kursky, M. D., and Vorobetz, Z. D. (1984*b*). Intracellular protein kinase and calcium inward currents in perfused neurones of the snail *Helix pomatia*. *Neuroscience* **11**, 263–7.

Douglas, W. W. and Taraskevich, P. S. (1980). Calcium component to action potentials in rat pars intermedia cells. *Journal of Physiology (London)* **309**, 623–30.

Dravid, A. R. and Hammerschlag, R. (1975). Axoplasmic transport of proteins *in vitro* in primary afferent neurons of frog spinal cord: Effect of Ca^{2+}-free incubation conditions. *Journal of Neurochemistry* **24**, 711–18.

Dunlap, K. (1977). Localization of calcium channels in *Paramecium caudatum*. *Journal of Physiology (London)* **271**, 119–33.

Dunlap, K. and Fischbach, G. D. (1978). Neurotransmitters decrease the calcium component of sensory neurone action potentials. *Nature* **276**, 837–9.

Dunlap, K. and Fischbach, G. D. (1981). Neurotransmitters decrease the calcium conductance activated by depolarization of embryonic chick sensory neurones. *Journal of Physiology (London)* **317**, 519–35.

Dupon, J.-L., Bossu, J.-L., and Feltz, A. (1986). Effect of internal calcium concentration on calcium currents in rat sensory neurones. *Pfluegers Archiv* **406**, 433–5.

Eckert, R. and Ewald, D. (1983*a*). Calcium tail currents in voltage-clamped intact nerve cell bodies of *Aplysia californica*. *Journal of Physiology (London)* **345**, 533–48.

Eckert, R. and Ewald, D. (1983*b*). Inactivation of calcium conductance characterized by tail current measurements in neurones of *Aplysia californica*. *Journal of Physiology (London)* **345**, 549–65.

Eckert, R. and Lux, H. D. (1976). A voltage-sensitive persistent calcium conductance in neuronal somata of *Helix*. *Journal of Physiology (London)* **254**, 129–51.

Eckert, R. and Tillotson, D. (1978). Outward-current component associated with intracellular Ca accumulation in aequorin-injected, voltage-clamped *Dorid* neurons. *Biophysical Journal* **21**, 178a.

Eckert, R. and Tillotson, D. (1981). Calcium-mediated inactivation of the calcium conductance in cesium-loaded giant neurones of *Aplysia californica*. *Journal of Physiology (London)* **314**, 265–80.

Eckert, R., Chad, J. E., and Kalman, D. (1986). Enzymatic regulation of calcium current in dialyzed and intact molluscan neurons. *Journal de Physiologie (Paris)* **81**, 318–24.

Ehrlich, B. E., Jacobson, A. R., Hinrichsen, R., Sayre, L. M., and Forte, M. A. (1988). *Paramecium* calcium channels are blocked by a family of calmodulin antagonists. *Proceedings of the National Academy of Sciences USA* **85**, 5718–22.

Eisenberg, M., Gresalfi, T., Riccio, T., and McLaughlin, S. (1979). Adsorption of monovalent cations to bilayer membranes containing negative phospholipids. *Biochemistry* **18**, 5213–23.

Ewald, D. and Eckert, R. (1983). Cyclic AMP enhances calcium-dependent potassium current in *Aplysia* neurons. *Cellular and Molecular Neurobiology* **3**, 345–54.

Ewald, D. and Levitan, I. B. (1984). Cyclic AMP-dependent protein phosphorylation modulates the activity of single potassium channels in isolated membrane patches and in reconstituted phospholipid bilayers. *Society for Neuroscience Abstracts* **10**, 145.

Ewald, D. A., Walker, M. W., Perney, T. M., Matthies, H. J. G., and Miller, R. J. (1988). Neurotransmitter modulation of calcium currents in rat sensory neurons. In *Calcium and ion channel modulation*, Proceedings of a symposium honouring R., Eckert, 26 February–1 March 1987, Los Angeles, California (ed. A. D. Grinnell, D. Armstrong, and M. B. Jackson), pp. 263–73. Plenum, New York.

Eyring, H., Lumry, R., and Woodbury, J. W. (1949). Some recent applications of modern rate theory to physiological systems. *Recent Chemical Progress* **10**, 100–14.

Fabiato, A. (1988). Mechanism and physiological relevance of calcium-induced release of calcium from cardiac sarcoplasmic reticulum: comparison with inositol

(1,4,5)-trisphosphate induced release of calcium. *Molecular and Cellular Cardiology* **2Q**, (4), 16.

Fasolato, C., Pandiella, A., Meldolesi, J., and Pozzan, T. (1988). Generation of inositol phosphates, cytosolic Ca^{++}, and ionic fluxes in PC12 cells treated with bradykinin. *Journal of Biological Chemistry* **263**, 17350–9.

Fatt, P. and Ginsborg, B. L. (1958). The ionic requirement for the production of action potentials in crustacean muscle fibres. *Journal of Physiology (London)* **142**, 516–43.

Fatt, P. and Katz, B. (1953). The electrical properties of crustacean muscle fibres. *Journal of Physiology (London)* **120**, 171–204.

Fedulova, S. A., Kostyuk, P. G., and Veselovsky, N. S. (1981). Calcium channels in the somatic membrane of the rat dorsal root ganglion neurons, effect of cAMP. *Brain Research* **214**, 210–14.

Fedulova, S. A., Kostyuk, P. G., and Veselovsky, N. S. (1985). Two types of calcium channels in the somatic membrane of newborn rat dorsal root ganglion neurons. *Journal of Physiology (London)* **359**, 431–46.

Fedulova, S. A., Kostyuk, P. G., and Veselovsky, N. S. (1986). Changes in ionic mechanisms of electrical excitability of the somatic membrane of rat's dorsal root ganglion neurons during ontogenesis. Correlation between inward current densities. *Neurophysiology (Kiev)* **18**, 820–7.

Fenwick, E. M., Marty, A., and Neher, E. (1982*a*). A patch-clamp study of bovine chromaffin cells and of their sensitivity to acetylcholine. *Journal of Physiology (London)* **331**, 577–97.

Fenwick, E. M., Marty, A., and Neher, E. (1982*b*). Sodium and calcium channels in bovine chromaffin cells. *Journal of Physiology (London)* **331**, 599–635.

Ferry, D. R. and Glossmann, H. (1982). Evidence for multiple receptor sites within the putative calcium channel. *Naunyn-Schmiedebergs Archives of Pharmacology* **320**, 41–4.

Ferry, D. R., Goll, A., and Glossmann, H. (1983). Differential labelling of putative skeletal muscle calcium channels by (^{3}H)-nifedipine, (^{3}H)-nitrendipine, (^{3}H)-nimodipine and (^{3}H)-PN200110. *Naunyn-Schmiedebergs Archives of Pharmacology* **323**, 276–7.

Field, A. C., Hill, C., and Lamb, G. D. (1988). Asymmetric charge movement and calcium currents in ventricular myocytes of neonatal rat. *Journal of Physiology* **406**, 277–97.

Fill, M. and Coronado, R. (1988). Ryanodine receptor channel of sarcoplasmic reticulum. *Trends in Neurosciences* **11**, 453–7.

Findlay, I. and Dunne, M. J. (1985). Voltage-activated Ca^{2+} currents in insulin-secreting cells. *FEBS Letters* **189**, 281–5.

Fink, L. A., Connor, J. A., and Kaczmarek, L. K. (1988). Inositol triphosphate releases intracellularly stored calcium and modulates ion channels in molluscan neurons. *Journal of Neuroscience* **8**, 2544–55.

Fischmeister, R. and Hartzell, H. C. (1986). Mechanism of action of actylcholine on calcium current in single cells from frog ventricle. *Journal of Physiology (London)* **376**, 183–202.

Fischmeister, R. and Vassort, G. (1981). The electrogenic Na–Ca exchange and the cardiac electrical activity. I. Simulation on Purkinje fibre action potential. *Journal of Physiology (Paris)* **77**, 705–9.

Fishman, M. C. and Spector, I. (1981). Potassium current suppression by quinidine reveals additional calcium currents in neuroblastoma cells. *Proceedings of the National Academy of Sciences USA* **78**, 5245–9.

Fiskum, G. and Lehninger, A. L. (1980). The mechanisms and regulation of mitochondrial Ca^{2+} transport. *Federation Proceedings* **39**, 2432–6.

Fleckenstein, A. (1977). Specific pharmacology of calcium in myocardium, cardiac pacemaker and vascular smooth muscle. *Annual Review of Pharmacology and Toxicology* **17**, 149–66.

Fleckenstein, A., Tritthart, H., Fleckenstein, B., Herbst, A., and Grun, G. Eine neue Gruppe kompetitiver Ca^{++}-antagonisten (Iproveratril, D-600, Prenylamin) mit starken Hemmeffekten auf die elektromechanische Kopellung in Warmblueter-Myokard. *Pfluegers Archiv* **307**, 25.

Forscher, P. and Oxford, G. S. (1984). Norepinephrine modulation of calcium channels in internally dialyzed sensory neurons. *Biophysical Journal* **45**, 181a.

Forscher, P., Oxford, G. S., and Schulz, D. (1986). Noradrenaline modulates calcium channels in avian dorsal root ganglion cells through tight receptor-channel coupling. *Journal of Physiology (London)* **379**, 131–44.

Fox, A. P. and Krasne, S. (1981). Two calcium currents in egg cells. *Biophysical Journal* **33**, (2), 145a.

Fox, A. P., Nowycky, M. C., and Tsien, R. W. (1984). Single Ca channels in dorsal root ganglion cells are shifted between modes of gating by the calcium agonist Bay K 8644. *Society for Neuroscience Abstracts* **10**, 526.

Fox, A. P., Nowycky, M. C., and Tsien, R. W. (1987*a*). Kinetic and pharmacological properties distinguishing three types of calcium currents in chick sensory neurones. *Journal of Physiology (London)* **394**, 149–72.

Fox, A. P., Nowycky, M. C., and Tsien, R. W. (1987*b*). Single-channel recordings of three types of calcium channels in chick sensory neurones. *Journal of Physiology (London)* **394**, 173–200.

Frankenhaeuser, B. (1957). The effect of calcium on the myelinated nerve fibre. *Journal of Physiology (London)* **137**, 245.

Frankenhaeuser, B. and Hodgkin, A. L. (1957). The action of calcium on the electrical properties of squid axons. *Journal of Physiology (London)* **137**, 218–24.

Freedman, S. B. and Miller, R. J. (1984). Effects of nitrendipine on voltage sensitive calcium channels in brain and neuronal cultured cells. In *Nitrendipine* (ed. A. Scriabine, S. Vanov, and K. Deck), pp. 79–90. Urban and Schwarzenberg, Baltimore.

Froehner, S. C. (1988). New insights into the molecular structure of the dihydropyridine-sensitive calcium channel. *Trends in Neurosciences* **11**, 90–2.

Fukuda, J. and Kameyama, M. (1979). Enhancement of Ca spikes in nerve cells of adult mammals during neurite growth in tissue culture. *Nature* **279**, 546–8.

Fukishima, Y. and Hagiwara, S. (1983). Voltage-gated Ca^{2+} channel in mouse myeloma cells. *Proceedings of the National Academy of Sciences USA* **80**, 2240–2.

Galizzi, J.-P., Qar, J., Fosset, M., Van Rentergham, C., and Lazdunski, M. (1987). Regulation of calcium channels in aortic muscle cells by protein kinase C activators (diacylglycerol and phorbol esters) and by peptides (vasopressin and bombesin) that stimulate phosphoinositide breakdown. *Journal of Biological Chemistry* **262**, 6947–50.

Geduldig, D. and Gruener, R. (1970). Voltage clamp of the *Aplysia* giant neurone: Early sodium and calcium currents. *Journal of Physiology (London)* **211**, 217–44.

Geduldig, D. and Junge, D. (1968). Sodium and calcium components of action potentials in the *Aplysia* giant neurone. *Journal of Physiology (London)* **199**, 347–65.

Geffen, L. B., Jessell, T. M., Cuello, A. C., and Iversen, L. L. (1976). Release of dopamine from dendrites in rat substantia nigra. *Nature* **260**, 258–60.

Gerasimov, V. D., Kostyuk, P. G., and Maisky, V. A. (1964). Excitability of giant nerve cells in different kinds of Gastropoda in Na-free solutions. *Bulletin of Experimental Biology (Moscow)* **58**, 3–7.

Gerschenfeld, H. M., Hammond, C., and Paupardin-Tritsch, D. (1986). Modulation of the calcium current of molluscan neurones by neurotransmitters. *Journal of Experimental Biology* **124**, 73–91.

Gilbert, D. L. and Ehrenstein, G. (1969). Effect of divalent cations on potassium conductance of squid axons: Determination of surface charge. *Biophysical Journal* **9**, 447–64.

Glass, D. B. and Krebs, E. G. (1980). Protein phosphorylation catalyzed by cyclic AMP-dependent and cyclic GMP-dependent protein kinases. *Annual Review of Pharmacology and Toxicology* **20**, 363–88.

Glossmann, H., *et al.* (1983*a*). Identification of voltage operated calcium channels by binding studies: Differentiation of subclasses of calcium antagonist drugs with ^{3}H-nimodipine radioligand binding. *Journal of Receptor Research* **3**, 177–90.

Glossmann, H., Linn, T., Rombush, M., and Ferry, D. R. (1983*b*). Temperature-dependent regulation of D-*cis*-(^{3}H)diltiazem-binding to Ca^{2+} channels by 1,4-dihydropyridine channel agonists and antagonists. *FEBS Letters* **160**, 226–32.

Goerke, K. and Pierau, F. K. (1980). Spike potentials and membrane properties of dorsal root ganglion cells in pigeons. *Pfluegers Archiv* **386**, 21–8.

Goll, A., Ferry, D. R., and Glossmann, H. (1984). Target size analysis and molecular properties of Ca^{2+} channels labelled with (^{3}H)verapamil. *European Journal of Neurochemistry* **141–**, 177–86.

Gonoi, T. and Hagesawa, S. (1988). Postnatal disappearance of transient calcium channels in mouse skeletal muscle: effects of denervation and culture. *Journal of Physiology (London)* **401**, 617–37.

Gorman, A. L. F. and Hermann, A. (1979). Internal effects of divalent cations on potassium permeability in molluscan neurones. *Journal of Physiology (London)* **296**, 393–410.

Gorman, A. L. F. and Thomas, M. V. (1978). Changes in the intracellular concentration of free calcium ions in a pacemaker neuron, measured with the metallochromic indicator dye arsenazo III. *Journal of Physiology (London)* **275**, 357–76.

Gorman, A. L. F., Hermann, A., and Thomas, M. V. (1981). Intracellular calcium and the control of neuronal pacemaker activity. *Federation Proceedings* **40**, 2233–9.

Gorman, A. L. F., Woolum, J. C., and Cornwell, H. C. (1982). Selectivity of the Ca^{2+}-activated and light-dependent K^{+} channels for monovalent cations. *Biophysical Journal* **38**, 319–22.

Gorman, A. L. F., Levy, S., Nsi, E., and Tillotson, D. (1984). Intracellular calcium measured with calcium-sensitive micro-electrodes and Arsenazo III in voltage-clamped *Aplysia* neurones. *Journal of Physiology (London)* **353**, 127–42.

Gould, R. J., Murphy, K. M. M., and Snyder, S. H. (1982). (^{3}H)nitrendipine-labeled calcium channels discriminate inorganic calcium agonists and antagonists. *Proceedings of the National Academy of Sciences USA* **79**, 3656–60.

Gouy, G. (1910). Sur la constitution de la charge électrique à la surface d-un électrolyte. *Journal de Physiologie (Paris)* **9**, 457–68.

Grafstein, B. and Forman, D. S. (1980). Intracellular transport in neurons. *Physiological Reviews* **60**, 1167–1283.

Gray, P. T. A. (1988). Oscillations of free cytosolic calcium evoked by cholinergic and catecholaminergic agonists in rat parotid acinar cells. *Journal of Physiology (London)* **406**, 35–53.

Gray, R. and Johnston, D. (1987). Noraderenaline and β-adrenoreceptor agonists increase activity of voltage-dependent calcium channels in hippocampal neurons. *Nature* **327**, 620–2.

Greenberg, D. A. and Marks, S. S. (1988). Cation interactions with putative NMDA receptor-gated channels labeled by (^{3}H)MK-801 in rat cerebral cortex. *Neuroscience Letters* **95**, 236–40.

Greenberg, D. A., Carpenter, C. L., and Messing, R. O. (1987). Lectin-induced enhancement of voltage-dependent calcium flux and calcium channel antagonist binding. *Journal of Neurochemistry* **48**, 888–94.

Greenfield, S., Chéramy, A., Leviel, V., and Glowinski, J. (1980). *In vivo* release of acetylcholinesterase in cat substantia nigra and caudate nucleus. *Nature* 284, 355–7.

Greengard, P. (1978). *Cyclic nucleotides, phosphorylated proteins and neuronal function.* Raven Press, New York.

Grinvald, A. and Farber, I. C. (1981). Optical recording of calcium action potentials from growth cones of cultured neurons with a laser microbeam. *Science* **212**, 1164–7.

Gross, R. A. and Macdonald, R. L. (1987). Dynorphin A selectively reduces a large transient (N-type) calcium current of mouse dorsal root ganglion neurons in cell culture. *Proceedings of the National Academy of Sciences USA* **84**, 5469–73.

Gross, R. A. and Macdonald, R. L. (1988). Reduction of the same calcium current component by A and C kinase: differential pertussis toxin sensitivity. *Neuroscience Letters* **88**, 50–6.

Gross, R. A. and Macdonald, R. L. (1989). Cyclic AMP selectively reduces the N-type calcium current component of mouse sensory neurons in culture by enhancing inactivation. *Journal of Neurophysiology* **61**, 97–105.

Grygorczyk, R. and Schwarz, W. (1983). Properties of the Ca^{2+}-activated K^+ conductance of human red cells as revealed by the patch-clamp techniques. *Cell Calcium* **4**, 499–510.

Grynkiewicz, G., Poenie, M., and Tsien, R. Y. (1985). A new generation of Ca^{2+} indicators with greatly improved fluorescence properties. *Journal of Biological Chemistry* **260**, 3440–50.

Haga, N., Forte, M., Ramanathan, R., Hennessey, T., Takahashi, M., and Kung, C. (1984). Characterization and purification of a soluble protein controlling Ca-channel activity in *Paramecium*. *Cell* **39**, 71–8.

Hagiwara, N., Irisawa, H., and Kameyama, M. (1988). Contribution of two types of calcium currents to the pacemaker potentials of rabbit sino-atrial node cells. *Journal of Physiology (London)* **395**, 233–53.

Hagiwara, S. (1973). Ca spike. *Advances in Biophysics* **4**, 71–102.

Hagiwara, S. and Byerly, L. (1983). The calcium channel. *Trends in Neurosciences* **6**, 189–93.

Hagiwara, S. and Miyazaki, S. (1977). Changes in excitability of the cell membrane during 'differentiation without cleavage' in the egg of the annelid, *Chaetopterus pergamentaceus*. *Journal of Physiology (London)* **272**, 197–216.

Hagiwara, S. and Naka, K. I. (1964). The initiation of spike potential in barnacle muscle fibres under low intracellular Ca^{++}. *Journal of General Physiology* **48**, 141–62.

Hagiwara, S. and Nakajima, S. (1966). Effects of the intracellular Ca ion concentration upon the excitability of the muscle fiber membrane of a barnacle. *Journal of General Physiology* **49**, 807–18.

Hagiwara, S. and Ohmori, H. (1982). Studies of calcium channels in rat clonal pituitary cells with patch electrode voltage clamp. *Journal of Physiology (London)* **331**, 231–52.

Hagiwara, S. and Ohmori, H. (1983). Studies of single calcium channel currents in rat clonal pituitary cells. *Journal of Physiology (London)* **336**, 649–61.

Hagiwara, S., Chichifu, S., and Naka, K. I. (1964). The effects of various ions on resting spike potentials of barnacle muscle fibers. *Journal of General Physiology* **48**, 165–79.

Hagiwara, S., Fukuda, J., and Eaton, D. C. (1974). Membrane currents carried by Ca, Sr and Ba in barnacle muscle fiber during voltage clamp. *Journal of General Physiology* **63**, 564–78.

Hagiwara, S., Ozawa, S., and Sand, O. (1975). Voltage clamp analysis of two inward current mechanisms in the egg cell membrane of a starfish. *Journal of General Physiology* **65**, 617–44.

Hamill, O. P., Marty, A., Neher, E., Sakmann, B., and Sigworth, R. J. (1981). Improved patch-clamp techniques for high-resolution current recording from cells and cell-free membrane patches. *Pfluegers Archiv* **391**, 85–100.

Hammerschlag, R. and Lavoie, P. A. (1979). Initiation of fast axonal transport: Involvement of calcium during transfer of proteins from Golgi apparatus to the transport system. *Neuroscience* **4**, 1195–201.

Hammerschlag, R., Dravid, A. R., and Chiu, A. Y. (1975). Mechanism of axonal transport: A proposed role for calcium ions. *Science* **188**, 273–5.

Hammerschlag, R., Bakhit, C., Chiu, A. Y., and Dravid, A. R. (1977). Role of calcium in the initiation of fast axonal transport of protein: Effects of divalent cations. *Journal of Neurobiology* **8**, 439–51.

Hammond, C., Paupardin-Tritsch, D., Nairn, A., Greengard, P., and Gerschenfeld, H. M. (1987). Cholecystokinin induces a decrease in Ca^{2+} current in snail neurons that appear to be mediated by protein kinase C. *Nature* **325**, 809–11.

Hammoudah, M. M. *et al.* (1981). Interaction of La^{2+} with phosphatidyl-serine vesicles. *Biochimica et Biophysica Acta* **645**, 102–14.

Harris, K. M., Kongsamut, S., and Miller, R. J. (1986). Protein kinase C mediated regulation of calcium channels in PC-12 pheochromocytoma cells. *Biochemical and Biophysical Research Communications* **134**, 1298–305.

Harris-Warrick, R. M. *et al.* (1988). An α_{40} subunit of a GTP-binding protein immunologically related to Go mediates a dopamine-induced decrease of Ca^{2+} current in snail neurons. *Neuron* **1**, 27–32.

Henček, M. and Zachar, J. (1977). Calcium currents and conductances in the muscle membrane of the crayfish. *Journal of Physiology (London)* **268**, 51–71.

Hennessey, T. M. and Kung, C. (1984). An anticalmodulin drug, W-7, inhibits the voltage-dependent calcium current in *Paramecium caudatum*. *Journal of Experimental Biology* **110**, 169–81.

Henquin, J.-C. and Meissner, H. P. (1983). Dibutyryl cyclic AMP triggers Ca^{2+} influx and Ca^{2+}-dependent electrical activity in pancreatic B cells. *Biochemical and Biophysical Research Communication* **112**, 614–20.

Henquin, J.-C., Bozem, M., Schmeer, W., and Nenquin, M. (1987). Distinct mechanism for two amplification systems of insulin release. *Biochemical Journal* **246**, 393–9.

Hermann, A. and Gorman, A. L. F. (1981*a*). Effects of 4-aminopyridine on potassium currents in a molluscan neuron. *Journal of General Physiology* **78**, 63–86.

Hermann, A. and Gorman, A. L. F. (1981*b*). Effects of tetraethylammonium on potassium currents in a molluscan neuron. *Journal of General Physiology* **78**, 87–110.

Hescheler, J. and Trautwein, W. (1988). Modification of L-type calcium current by intracellularly applied trypsin in guinea-pig ventricular myocytes. *Journal of Physiology (London)* **404**, 259–74.

Hescheler, J., Pelzer, D., Trube, G., and Trautwein, W. (1982). Does the organic calcium channel blocker D-600 act from inside or outside on the cardiac cell membrane? *Pfluegers Archiv* **393**, 287–91.

Hescheler, J., Kameyama, M., and Trautwein, W. (1986). On the mechanism of muscarinic inhibition of the cardiac Ca current. *Pfluegers Archiv* **407**, 182–9.

Hescheler, J., Rosenthal, W., Trautwein, W., and Schultz, G. (1987*a*). The GTP-binding protein, Go, regulates neuronal calcium channels. *Nature* **325**, 445–7.

Hescheler, J., Tang, M., Jastorff, B., and Trautwein, W. (1987*b*). On the mechanism of histamine induced enhancement of the cardiac Ca^{2+} current. *Pfluegers Archiv* **410**, 23–9.

Hess, P. and Tsien, R. W. (1984). Mechanism of ion permeation through calcium channels. *Nature* **309**, 453–6.

Hess, P., Lansman, J. B., and Tsien, R. W. (1984). Modulation of single calcium channels by the calcium agonist Bay K 8644. *Biophysical Journal* **45**, 394.

Hess, P., Lansman, J. B., and Tsien, R. W. (1986). Calcium channel selectivity for divalent and monovalent cations. Voltage and concentration dependence of single channel current in ventricular heart cells. *Journal of General Physiology* **88**, 293–319.

Hille, B. (1968). Charges and potentials at the nerve surface: Divalent ions and pH. *Journal of General Physiology* **51**, 221–36.

Hille, B. (1975). Ionic selectivity, saturation and block in sodium channels: A four-barrier model. *Journal of General Physiology* **66**, 535–60.

Hille, B., Woodhull, A. M., and Shapiro, B. I. (1975). Negative surface charge near sodium channels of nerve: Divalent ions, monovalent ions and pH. *Philosophical Transactions of the Royal Society of London* **B270**, 301–12.

Hinke, M. T., Rinaldi, M. L., Capony, J. P., and Demaille, J. G. (1984). Ca^{2+}-flux modulation by calciductin phosphorylation in cardiac sarcolemma. *Membrane Biochemistry* **5**, 109–17.

Hiriart, M. and Matteson, D. R. (1988). Na-channels and two types of Ca-channels in rat pancreatic B cells identified with the reverse hemolytic plaque assay. *Journal of General Physiology* **91**, 617–39.

Hirning, L. D., *et al.* (1988). Dominant role of N-type Ca^{2+} channels in evoked release of norepinephrine from sympathetic neurons. *Science* **239**, 57–61.

Hirst, G. D. S. and Spence, I. (1973). Calcium action potentials in mammalian peripheral neurones. *Nature, New Biology* **243**, 54–6.

Hochner, B., Parnas, H., and Parnas, I. (1989). Membrane depolarization evokes neurotransmitter release in the absence of calcium entry. *Nature* **342**, 433–5.

Hodgkin, A. L. and Huxley, A. F. (1952). A quantitative description of membrane current and its application to conduction and excitation in nerve. *Journal of Physiology (London)* **117**, 500–44.

Hodgkin, A. L. and Keynes, R. D. (1957). Movements of labelled calcium in squid giant axons. *Journal of Physiology (London)* **138**, 253–81.

Hofmeier, G. and Lux, H. D. (1978). Inversely related behaviour of potassium and calcium permeability during activation of calcium-dependent outward currents in voltage-clamped snail neurones. *Journal of Physiology (London)* **287**, 28P–29P.

Holl, R. W., Thorner, M. O., Mandell, G. L., Sullivan, J. A., Sinha, Y. N., and Leong, D. A. (1988). Spontaneous oscillations of intracellular calcium and growth hormone secretion. *Journal of Biological Chemistry* **263**, 9682–5.

Holman, M. E. (1958). Membrane potentials recorded with high-resistance micro-electrodes and the effects of changes in ionic environment on the electrical and mechanical activity of the smooth muscle of the taenia coli of the guinea-pig. *Journal of Physiology (London)* **141**, 464–88.

Horn, J. P. and McAfee, D. A. (1979). Norepinephrine inhibits calcium-dependent potentials in rat sympathetic neurons. *Science* **204**, 1233–5.

Horn, J. P. and McAfee, D. A. (1980). Alpha-adrenergic inhibition of calcium-dependent potentials in rat sympathetic neurones. *Journal of Physiology (London)* **301**, 191–204.

Hosey, M. M., Borsotto, M., and Lazdunski, M. (1986). Phosphorylation and dephosphorylation of dihydropyridine-sensitive voltage-dependent Ca^{2+} channel in skeletal muscle membranes by cAMP- and Ca^{2+}-dependent processes. *Proceedings of the National Academy of Sciences USA* **83**, 3733–7.

Huttner, W. B., Schiebler, W., Greengard, P., and DeCamilli, P. (1983). Synapsin I (protein I), a nerve terminal-specific phosphoprotein. III. Its association with synaptic vesicles studied in a highly purified synaptic vesicle preparation. *Journal of Cell Biology* **96**, 1374–88.

Hymel, L., Schindler, H., Inui, M., and Fleischer, S. (1988). Reconstitution of purified cardiac muscle calcium release channel (ryanodine receptor) in planar bilayers. *Biochemical and Biophysical Research Communications* **152**, 308–14.

Ikeda, S. R. and Schofield, G. G. (1989). Somatostatin blocks calcium current in rat sympathetic ganglion neurones. *Journal of Physiology (London)* **409**, 221–40.

Ikemoto, Y., Mitsuiye, T., and Ishizuka, S. (1986). Reduction of the voltage-dependent calcium current in *Aplysia* neurons by pentobarbital. *Cellular and Molecular Neurobiology* **6**, 293–5.

Imagawa, T., Leung, A. T., and Campbell, K. P. (1987*a*). Phosphorylation of the 1,4-dihydropyridine receptor of the voltage-dependent Ca^{2+} channel by an intrinsic protein kinase in isolated triads from rabbit skeletal muscle. *Journal of Biological Chemistry* **262**, 8333–9.

Imagawa, T., Smith, J. S., Coronado, R., and Campbell, K. P. (1987*b*). Purified ryanodine receptor from skeletal muscle sarcoplasmic reticulum is the Ca^{2+}-permeable pore of the calcium release channel. *Journal of Biological Chemistry* **262**, 16636-43.

Isenberg, G. and Kloeckner, U. (1982). Calcium currents of isolated bovine ventricular myocytes are fast and of large amplitude. *Pfluegers Archiv* **395**, 30–41.

Iwasaki, S. and Satow, Y. (1971). Sodium- and calcium-dependent spike potentials in the secretory neuron soma of the X-organ in the crayfish. *Journal of General Physiology* **57**, 216–38.

Iwatsuki, N. and Petersen, O. H. (1978). Intracellular Ca^{2+} injection causes membrane hyperpolarization and conductance increase in lacrimal acinar cells. *Pfluegers Archiv* **377**, 185–7.

Jacobs, K. H., Aktories, K., and Schultz, G. (1984). Inhibitory modulation of adenylate cyclase. *IUPHAR 9th International Congress of Pharmacology*, Abstracts L., pp. S7–2. Macmillan, London.

Jaffé, E. H. and Cuello, A. C. (1980). Release of gamma-aminobutyrate from the external plexiform layer of the rat olfactory bulb: possible dendritic involvement. *Neuroscience* **5**, 1859–69.

Jaffe, L. A., Gould-Somero, M., and Holland, L. (1979). Ionic mechanism of the fertilization potential of the marine worm, *Urechis caupo* (Echiura). *Journal of General Physiology* **73**, 469–92.

Janis, R. A. and Triggle, D. J. (1983). New developments in Ca^{2+} channel antagonists. *Journal of Medicinal Chemistry* **26**, 775–85.

Janis, R. A. and Triggle, D. J. (1984). 1,4-dihydropyridine Ca^{2+} channel antagonists and activators: A comparison of binding characteristics with pharmacology. *Drug Development Research* **4**, 257–74.

Janis, R. A. *et al.* (1988). Isolation and characterization of a fraction from brain that inhibits 1,4,-(^{3}H)dihydropyridine binding and L-type calcium channel current. *FEBS Letters* **239**, 233–6.

Jeng, A. Y. and Shamoo, A. E. (1980). Isolation of a Ca^{2+}-carrier from calf heart inner mitochondrial membrane. *Journal of Biological Chemistry* **255**, 6897–903.

Johnson, J. D. (1984). A calmodulin-like Ca^{2+}-receptor in the Ca^{2+} channel. *Biophysical Journal* **45**, 134–6.

Johnson, J. D., Wittenauer, L. A., and Nathan, R. D. (1983). Calmodulin, Ca^{2+}-antagonists and Ca^{2+}-transporters in nerve and muscle. *Journal of Neural Transmission* Suppl. 18, 97–111.

Johnson, M. and Mitchell, R. (1989). Phorbol ester-induced secretion of GH but not LH involves activation of L-type calcium channels. *Journal of Endocrinology* **121**, 49.

Junge, D. (1967). Multi-ionic action potentials in molluscan giant neurones. *Nature* **215**, 546–8.

Kaczmarek, L. K., *et al.* (1980). Microinjection of catalytic subunit of cyclic AMP-dependent protein kinase enhances calcium action potentials of bag cell neurons in cell culture. *Proceedings of the National Academy of Sciences USA* **77**, 7487–91.

Kalman, D., O'Lague, P. H., Erxleben, C., and Armstrong, D. L. (1988). Calcium-dependent inactivation of the dihydropyridine-sensitive calcium channels in GH3 cells. *Journal of Physiology (London)* **92**, 531–48.

Kameyama, M., Kameyama, A., Nakayama, T., and Kaibara, M. (1988). Tissue extract recovers cardiac calcium channels from "run-down". *Pfluegers Archiv* **412**, 328–30.

Kaneda, M., Nakamura, H., and Akaike, N. (1988). Mechanical and enzymatic isolation of mammalian CNS neurones. *Neuroscience Research* **5**, 299–315.

Kaneko, A., Pinto, L. H., and Tachibana, M. (1989). Transient calcium current of retinal bipolar cells of the mouse. *Journal of General Physiology* **410**, 613–29.

Kaneko, S. and Nomura, Y. (1987). Cyclic AMP facilitates slow-inactivating Ca^{2+} channel currents expressed by *Xenopus* oocyte after injection of rat brain mRNA. *Neuroscience Letters* **83**, 123–7.

Kao, C. Y. and McCullough, J. R. (1975). Ionic currents in the uterine smooth muscle. *Journal of Physiology (London)* **246**, 1–36.

Kass, R. S. and Arena, J. P. (1989). Influence of pHo on calcium channel block by amlodipine, a charged dihydropyridine compound. Implications for location of the dihydropyridine receptor. *Journal of General Physiology* **93**, 1109–27.

Kass, R. S. and Sanguinetti, M. C. (1984). Inactivation of calcium channel current in the calf cardiac Purkinje fiber. Evidence for voltage- and calcium-mediated mechanisms. *Journal of General Physiology* **84**, 705–26.

Katz, B. and Miledi, R. (1967*a*). The timing of calcium action during neuromuscular transmission. *Journal of Physiology (London)* **189**, 535–44.

Katz, B. and Miledi, R. (1967*b*). A study of synaptic transmission in the absence of nerve impulses. *Journal of Physiology (London)* **192**, 407–36.

Katz, B. and Miledi, R. (1970). Further study of the role of calcium in synaptic transmission. *Journal of Physiology (London)* **207**, 789–802.

Kay, A. R. and Wong, R. K. S. (1987). Calcium current activation kinetics in isolated pyramidal neurones of the CAl region of the mature guinea-pig hippocampus. *Journal of Physiology (London)* **392**, 603–16.

Kazazoglou, T., Schmid, A., Renaud, J. F., and Lazdunski, M. (1983). Ontogenetic appearance of Ca^{2+} channels characterized as binding sites for nitrendipine during development of nervous, skeletal and cardiac muscle systems in the rat. *FEBS Letters* **161**, 75–80.

Kazda, S., Garthoff, B., and Knorr, A. (1983). Nitrendipine and other calcium entry blockers (calcium antagonists) in hypertension. *Federation Proceedings* **42**, 196–200.

Keynes, R. D. and Lewis, P. R. (1956). The intracellular calcium content of some invertebrate nerves. *Journal of Physiology (London)* **134**, 399–407.

Keynes, R. D. and Rojas, E. (1974). Kinetics and steady-state properties of the charged system controlling sodium conductance in the squid giant axon. *Journal of Physiology (London)* **239**, 393–434.

Keynes, R. D., Rojas, E., Taylor, R. E., and Vergara, J. (1973). Calcium and potassium systems of a giant barnacle muscle fibre under membrane potential control. *Journal of Physiology (London)* **229**, 409–55.

Khodorov, B. I. (1979). Some aspects of the pharmacology of sodium channels in nerve membrane: Process of inactivation. *Biochemical Pharmacology* **28**, 1451–9.

Klein, M. and Kandel, E. R. (1978). Presynaptic modulation of voltage-dependent Ca^{2+} current: Mechanism for behavioural sensitization in *Aplysia californica*. *Proceedings of the National Academy of Sciences USA* **75**, 3512–16.

Klein, M. and Kandel, E. R. (1980). Mechanism of calcium modulation underlying presynaptic facilitation and behavioural sensitization in *Aplysia*. *Proceedings of the National Academy of Sciences USA* **77**, 6912–16.

Kleinhaus, A. L. and Prichard, J. W. (1975). Calcium dependent action potentials produced in leech Retzius cells by tetraethylammonium chloride. *Journal of Physiology (London)* **246**, 351–69.

Kley, N., Loeffler, J.-P., Pittius, C. W., and Hollt, V. (1987). Involvement of ion channels in the induction of proenkephalin A gene expression by nicotine and cAMP in bovine chromaffin cells. *Journal of Biological Chemistry* **262**, 4083–9.

Kohlhardt, M., Bauer, B., Krause, H., and Fleckenstein, A. (1972*a*). New selective inhibitors of the transmembrane Ca conductivity in mammalian myocardial fibres: studies with the voltage clamp technique. *Experientia* **28**, 288–9.

Kohlhardt, M., Bauer, B., Krause, H., and Fleckenstein, A. (1972*b*). Differentiation of the transmembrane Na and Ca channels in mammalian cardiac fibres by the use of specific inhibitors. *Pfluegers Archiv* **335**, 309–22.

Kohlhardt, M., Bauer, B., Krause, H., and Fleckenstein, A. (1973). Selective inhibition of the transmembrane Ca conductivity of mammalian myocardial fibres by Ni, Co and Mn ions. *Pfluegers Archiv* **338**, 115–23.

Koketsu, K. and Akasu, T. (1982). Modulation of the slow inward Ca^{2+} current by adrenaline in bullfrog sympathetic ganglion cells. *Japanese Journal of Physiology* **32**, 137–40.

Koketsu, K. and Nishi, S. (1969). Calcium and action potentials of bullfrog sympathetic ganglion cells. *Journal of General Physiology* **53**, 608–23.

Kongsamut, S., Freedman, S. B., and Miller, R. J. (1985). Dihydropyridine sensitive calcium channels in a smooth muscle cell line. *Biochemical and Biophysical Research Communications* **127**, 71–9.

Konnerth, A., Lux, H. D., and Morad, M. (1987). Proton-induced transformation of calcium channel in chick dorsal root ganglion cells. *Journal of Physiology (London)* **386**, 603–33.

Kononenko, N. I. (1984). Synaptic regulation of calcium current in an identified RPa1 neuron of *Helix pomatia*. *Doklady Akademii Nauk SSSR* **275**, 231–4.

Kononenko, N. I. and Mironov, S. L. (1980). The effect of intracellular injection of cyclic adenosine monophosphate on electrical characteristics of identified snail neurons. *Neurophysiology (Kiev)* **12**, 517–25.

Kononenko, N. I. and Shcherbatko, A. D. (1985*a*). The effect of an increase of the intracellular calcium concentration on transmembrane current evoked by iontophoretic cAMP injection into snail neurons. *Neurophysiology (Kiev)* **17**, 78–84.

Kononenko, N. I. and Shcherbatko, A. D. (1985*b*). Influence of serotonin on the inward calcium current in *Helix* neurones. *Doklady Academii Nauk SSSR* **281**, 1494–7.

Kononenko, N. I. and Shcherbatko, A. D. (1988*a*). Effect of iontophoretic injection of AMP and cAMP on the calcium current in dialyzed snail neurons. *Neurophysiology (Kiev)* **20**, 769–76.

Kononenko, N. I. and Shcherbatko, A. D. (1988*b*). Furosemide effect on the inward calcium current in snail neurons. *Biological Membranes (Moscow)* **5**, 1168–77.

Kononenko, N. I., Kostyuk, P. G., and Shcherbatko, A. D. (1983). The effect of intracellular cAMP injections on stationary membrane conductance and voltage-

and time-dependent ionic currents in identified snail neurons. *Brain Research* **268**, 321–8.

Kononenko, N. I., Kostyuk, P. G., and Shcherbatko, A. D. (1986). Properties of cAMP-induced transmembrane current in mollusc neurons. *Brain Research* **376**, 239–45.

Kostyuk, P. G. (1981). Intracellular perfusion. In *Techniques in the life sciences: Techniques in cellular physiology*, (ed. P. F. Baker), P1/1, pt. 1, P110/1–P110/7. Elsevier/North-Holland, Amsterdam.

Kostyuk, P. G. (1982). Intracellular perfusion. *Annual Review of Neuroscience* **5**, 107–20.

Kostyuk, P. G. and Krishtal, O. A. (1977*a*). Separation of sodium and calcium currents in the somatic membrane of mollusc neurones. *Journal of Physiology (London)* **270**, 545–68.

Kostyuk, P. G. and Krishtal, O. A. (1977*b*). Effects of calcium and calcium-chelating agents on the inward and outward currents in the membrane of mollusc neurones. *Journal of Physiology (London)* **270**, 569–80.

Kostyuk, P. G. and Krishtal, O. A. (1984). Intracellular perfusion of excitable cells. In *Methods in neuroscience*, Vol. 5, IBRO Handbook Series. Wiley & Sons, Chichester.

Kostyuk, P. G. and Mironov, S. L. (1982). Theoretical description of calcium channels in the neuronal membrane. *General Physiology and Biophysics* **1**, 289–305.

Kostyuk, P. G. and Mironov, S. L. (1986). Some predictions concerning the calcium channel model with different conformational states. *General Physiology and Biophysics* **6**, 649–59.

Kostyuk, P. G. and Shirokov, R. E. (1989). Deactivation kinetics of different components of calcium inward current in the membrane of mice sensory neurones. *Journal of Physiology (London)* **409**, 343–55.

Kostyuk, P. G. and Shuba, Ya. M. (1982). A study of monovalent cation selectivity of calcium EDTA-modified channels. *Neurophysiology (Kiev)* **14**, 491–8.

Kostyuk, P. G., Krishtal, O. A., and Doroshenko, P. A. (1974*a*). Calcium currents in snail neurones. I. Identification of calcium current. *Pfluegers Archiv* **348**, 83–93.

Kostyuk, P. G., Krishtal, O. A., and Doroshenko, P. A. (1974*b*). Calcium currents in snail neurones. II. The effect of external calcium concentration on the calcium inward current. *Pfluegers Archiv* **348**, 95–104.

Kostyuk, P. G., Krishtal, O. A., and Pidoplichko, V. I. (1975). Effect of internal fluoride and phosphate on membrane currents during intracellular dialysis of nerve cells. *Nature* **257**, 691–3.

Kostyuk, P. G., Krishtal, O. A., and Pidoplichko, V. I. (1977). Asymmetrical displacement currents in nerve cell membrane and effect of internal fluoride. *Nature* **267**, 70–2.

Kostyuk, P. G., Krishtal, O. A., and Pidoplichko, V. I. (1978). Ionic currents in the neuroblastoma cell membrane. *Neuroscience* **3**, 327–32.

Kostyuk, P. G., Krishtal, O. A., Pidoplichko, V. I., and Shakhovalov, Yu. A. (1979). Kinetics of calcium inward current activation. *Journal of General Physiology* **73**, 675–7.

Kostyuk, P. G., Doroshenko, P. A., and Tsyndrenko, A. Ya. (1980*a*). Calcium-dependent potassium conductance studied on internally dialysed nerve cells. *Neuroscience* **5**, 2187–92.

Kostyuk, P. G., Krishtal, O. A., and Pidoplichko, V. I. (1980*b*). Gating current of calcium channels in the neuronal membrane. *Bioelectrochemistry and Bioenergetics* **7**, 421–31.

Kostyuk, P. G., Krishtal, O. A., and Pidoplichko, V. I. (1981*a*). Intracellular perfusion. *Journal of Neuroscience Methods* **4**, 201–10.

Kostyuk, P. G., Krishtal, O. A., and Pidoplichko, V. I. (1981*b*). Calcium inward current and related charge movements in the membrane of snail neurones. *Journal of Physiology (London)* **310**, 403–21.

Kostyuk, P. G., Fedulova, S. A., Mironov, S. L., and Veselovsky, N. S. (1981*c*). Calcium currents in mammalian neurons. In *The mechanism of gated calcium transport across biological membranes*, (ed. S. T. Ohnishi and M. Endo), pp. 35–42. Academic Press, New York.

Kostyuk, P. G., Mironov, S. L., and Doroshenko, P. A. (1981*d*). The use of a three-barrier model for description of energy profile of the calcium channel in the membrane of mollusc neurones. *Neurophysiology (Kiev)* **13**, 322–31.

Kostyuk, P. G., Veselovsky, N. S., and Fedulova, S. A. (1981*e*). Ionic currents in the somatic membrane of rat dorsal root ganglion neurons. II. Calcium currents. *Neuroscience* **6**, 243–7.

Kostyuk, P. G., Veselovsky, N. S., and Tsyndrenko, A. Y. (1981*f*). Ionic currents in the somatic membrane of rat dorsal root ganglion neurons. I. Sodium currents. *Neuroscience* **6**, 2423–30.

Kostyuk, P. G., Mironov, S. L., and Doroshenko, P. A. (1982*a*). Energy profile of the calcium channel in the membrane of mollusc neurons. *Journal of Membrane Biology* **70**, 181–9.

Kostyuk, P. G., Mironov, S. L., Doroshenko, P. A., and Ponomarev, V. N. (1982*b*). Surface charges on outer side of mollusc neuron membrane. *Journal of Membrane Biology* **70**, 171–9.

Kostyuk, P. G., Mironov, S. L., and Shuba, Ya. M. (1983). Two ion-selecting filters in the calcium channel of the somatic membrane of mollusc neurons. *Journal of Membrane Biology* **76**, 83–93.

Kostyuk, P. G., Doroshenko, P. A., and Martynyuk, A. E. (1985*a*). Fast decrease of the current carried by barium ions through calcium channels in the somatic membrane of mollusc neurons. *Pfluegers Archiv* **404**, 88–90.

Kostyuk, P. G., Savchenko, A. N., and Shuba, Ya. M. (1985*b*). Barium currents through single calcium channels in the nerve cell membrane. *Doklady Akademii Nauk SSSR* **280**, 1459–62.

Kostyuk, P. G., Fedulova, S. A., and Veselovsky, N. S. (1986). Changes in ionic mechanisms of electrical excitability of the somatic membrane of rat dorsal root ganglion neurons during ontogenesis. Distribution of ionic channels of inward current. *Neurophysiology (Kiev)* **18**, 813–20.

Kostyuk, P. G., Shuba, Ya. M., and Savchenko, A. N. (1988*a*). Three types of calcium channels in the membrane of mouse sensory neurons. *Pfluegers Archiv* **411**, 661–9.

Kostyuk, P. G., Shuba, Ja. M., Savchenko, A. N., and Teslenko, V. I. (1988*b*). Kinetic characteristics of different calcium channels in the neuronal membrane. In *The calcium channel: Structure, function and implications*, (ed. Morad, M., Naylor, W., Kazda, S. and Schramm, M.), pp. 442–64. Springer-Verlag, Berlin.

Kostyuk, P. G., Mironov, S. L., Tepikin, A. V., and Belan, P. V. (1989*a*). Cytoplasmic free Ca in isolated snail neurons as revealed by fluorescent probe fura-2: mechanisms of Ca recovery after Ca load and Ca release from intracellular sites. *Journal of Membrane Biology* **110**, 11–18.

Kostyuk, P. G., Shuba, Ya. M., and Teslenko, V. I. (1989*b*). Activation kinetics of single high-threshold calcium channels in the membrane of sensory neurons from mouse embryos. *Journal of Membrane Biology* **110**, 29–38.

Kostyuk, P. G., Luk'yanetz, E. A., and Doroshenko, P. A. (1990). Studies of the cAMP influence on calcium currents in mollusc neurones possessing different sensitivity of their calcium conduction to serotonin action. *Neurophysiology (Kiev)* **22**, 605–12.

Kozuka, M. and Takahashi, K. (1982). Changes in holding and ion-channel currents during activation of an ascidian egg under voltage clamp. *Journal of Physiology (London)* **323**, 267–86.

Krishtal, O. A. and Magura, I. S. (1970). Calcium ions as inward current carriers in mollusc neurones. *Comparative Biochemistry and Physiology* **35**, 857–66.

Krishtal, O. A. and Pidoplichko, V. I. (1981). A receptor for protons in the membrane of sensory neurons may participate in nociception. *Neuroscience* **6**, 2599–601.

Krishtal, O. A., Pidoplichko, V. I., and Shakhovalov, Yu. A. (1980). Properties of single calcium channels in the neural membrane. *Bioelectrochemistry and Bioenergetics* **7**, 195–207.

Krishtal, O. A., Pidoplichko, V. I., and Shakhovalov, Yu. A. (1981). Conductance of the calcium channel in the membrane of snail neurones. *Journal of Physiology (London)* **310**, 423–34.

Krizanova, O., Hudecova, S., Orlicky, J., and Zachar, J. (1989). Is Ca^{2+} antagonists binding protein from cytosolic fraction the precursor of alpha 1-subunit of Ca^{++} channel? *General Physiology and Biophysics* **8**, 99–111.

Kumamoto, M. and Horn, L. (1970). Voltage clamping of smooth muscles from taenia coli. *Microvascular Research* **2**, 188–201.

Kunze, D. L., Lux, H. D., and Brown, A. M. (1984). Inactivation of single channel Ca currents. *Biophysical Journal* **45**, 179a.

Lacerda, A. E., Rampe, D., and Brown, A. M. (1988). Effects of protein kinase C activators on cardiac Ca^{2+} channels. *Nature* **335**, 249–51.

Lai, F. A., Anderson, K., Rousseau, E., Liu, Q.-Y., and Meissner, G. (1988). Evidence for a Ca^{2+} channel within the ryanodine receptor complex from cardiac sarcoplasmic reticulum. *Biochemical and Biophysical Research Communications* **151**, 441–9.

Landy, P. M. and Frew, R. (1988). Evidence of ω-conotoxin GVIA-sensitive Ca^{++} channels in mammalian peripheral nerve terminals. *European Journal of Pharmacology* **156**, 325–30.

Lavoie, P. A., Bolen, F., and Hammerschlag, R. (1979). Divalent cation specificity of the calcium requirement for fast transport of proteins in axons of desheated nerves. *Journal of Neurochemistry* **32**, 1745–51.

Lazarev, P. P. (1923). *Ionic theory of excitation*. State Publishing House, Moscow.

Lee, E., Lee, K. S., Noble, D., and Spindler, A. (1984). A new, very slow inward Ca current in single ventricular cells of adult guinea-pig. *Journal of Physiology (London)* **346**, 75P.

Lee, K. S. and Tsien, R. W. (1982). Reversal of current through calcium channels in dialysed single heart cells. *Nature* **297**, 498–501.

Lee, K. S., Akaike, N., and Brown, A. M. (1978). Properties of internally perfused, voltage-clamped, isolated nerve cell bodies. *Journal of General Physiology* **71**, 489–507.

Lee, K. S., Marban, E., and Tsien, R. W. (1985). Inactivation of calcium channels in mammalian heart cells: joint dependence on membrane potential and intracellular calcium. *Journal of Physiology (London)* **364**, 395–411.

Lemos, J. R. and Nowycky, M. C. (1989). Two types of calcium channels coexist in peptide-releasing vertebrate nerve terminals. *Neuron* **2**, 1419–26.

Levitan, E. S. and Levitan, I. B. (1988). Serotonin acting via cyclic AMP enhances both the hyperpolarizing and depolarizing phases of bursting pacemaker activity in the *Aplysia* neuron R15. *Journal of Neuroscience* **8**, 1152–61.

Levitan, I. B. (1985). Phosphorylation of ion channels. *Journal of Membrane Biology* **87**, 177–90.

Levitan, I. B. and Norman, J. (1980). Different effects of cAMP and cGMP derivatives on the activity of an identified neuron: Biochemical and electrophysiological analysis. *Brain Research* **187**, 415–29.

Lew, V. L. and Ferreira, H. G. (1976). Variable Ca sensitivity of a K-selective channel in intact red-cell membranes. *Nature* **263**, 336–8.

Lewis, D. L., Goodman, M. B., St John, P. A., and Barker, J. L. (1988). Calcium currents and fura-2 signals in fluorescence-activated cell sorted lactotrophs and somatotrophs of rat anterior pituitary. *Endocrinology* **123**, 611–21.

Lipscombe, D., *et al.* (1988*a*). Imaging of cytosolic Ca^{2+} transients arising from Ca^{2+} stores and Ca^{2+} channels in sympathetic neurons. *Neuron* **1**, 355–65.

Lipscombe, D., Madison, D. V., Poenie, M., Reuter, H., Tsien, R. Y., and Tsien, R. W. (1988*b*). Spatial distribution of calcium channels and cytosolic calcium transients in growth cones and cell bodies of sympathetic neurons. *Proceedings of the National Academy of Sciences USA* **85**, 2398–402.

Lishko, V. K., Zhukareva, V. A., and Malysheva, M. K. (1987). Purification and properties of the TTX-sensitive protein from bovine brain soluble fraction. In *Receptors and ion channels*, (ed. Yu. A. Ovchinnikov and F. Hucho), pp. 187–92. Walter de Gruyter, Berlin.

Llinas, R. (1979). The role of calcium in neuronal function. In *The neurosciences: Fourth study program*, (ed. F. O. Schmitt and F. G. Worden), pp. 555–71. MIT Press, Cambridge, Mass.

Llinas, R. and Hess, R. (1976). Tetrodotoxin-resistant dendritic spikes in avian Purkinje cells. *Proceedings of the National Academy of Sciences USA* **73**, 2520–3.

Llinas, R. and Nicholson, C. (1975). Calcium role in depolarization-secretion coupling: An aequorin study in squid giant synapse. *Proceedings of the National Academy of Sciences USA* **72**, 187–90.

Llinas, R. and Sugimori, M. (1980*a*). Electrophysiological properties of *in vitro* Purkinje cell somata in mammalian cerebellar slices. *Journal of Physiology (London)* **305**, 171–95.

Llinas, R. and Sugimori, M. (1980*b*). Electrophysiological properties of *in vitro* Purkinje cell dendrites in mammalian cerebeller slices. *Journal of Physiology (London)* **305**, 197–213.

Llinas, R., Blinks, J. R., and Nicholson, C. (1972). Calcium transient in presynaptic terminal of squid giant synapse: Detection with aequorin. *Science* **176**, 1127–9.

Llinas, R., Steinberg, I. Z., and Walton, K. (1976). Presynaptic calcium currents and their relation to synaptic transmission: Voltage-clamp study in squid giant synapse and theoretical model for the calcium gate. *Proceedings of the National Academy of Sciences USA* **73**, 2918–22.

Llinas, R., Steinberg, I. Z., and Walton, K. (1981*a*). Presynaptic calcium currents in squid giant synapse. *Biophysical Journal* **33**, 289–322.

Llinas, R., Steinberg, I. Z., and Walton, K. (1981*b*). Relationship between presynaptic calcium current and postsynaptic potential in squid giant synapse. *Biophysical Journal* **33**, 323–52.

Llinas, R., Greenfield, S. A., and Jahnsen, H. J. (1984). Electrophysiology of pars compacta cells in the *in vitro* substantia nigra—a possible mechanism for dendritic release. *Brain Research* **294**, 127–32.

Loeb, J. (1906). *The dynamics of living matter*, p. 134. New York.

Loirand, G., Pacaud, P., Mironneau, C., and Mironneau, J. (1986). Evidence for two distinct calcium channels in rat vascular smooth muscle cells in short-term primary culture. *Pfluegers Archiv* **407**, 566–8.

Lombet, A., Kazazoglou, T., Delpont, E., Renaud, J. F., and Lazdunski, M. (1983). Ontogenetic appearance of Na^+ channels characterized as high affinity sites for TTX during development of the rat nervous and skeletal muscle systems. *Biochemical and Biophysical Research Communications* **110**, 894–901.

Lorentz, M., Hedlund, B., and Arhem, P. (1988). Morphine activates calcium channels in cloned mouse neuroblastoma cell line. *Brain Research* **445**, 157–9.

Lotshaw, D. P. and Levitan, I. B. (1988). Reciprocal modulation of calcium current by serotonin and dopamine in the identified *Aplysia* neuron R15. *Brain Research* **439**, 64–76.

Lozinskaya, I. M. and Shuba, Ya. M. (1989). Low- and high-threshold calcium current in the membrane of cultivated *Helix pomatia* neurons. *Neurophysiology (Kiev)* **21**, 127–9.

Lüttgau, H. C. and Niedergerke, R. (1958). The antagonism between Ca and Na ions in the frog's heart. *Journal of Physiology (London)* **143**, 486–505.

Lux, H. D. and Brown, A. M. (1984). Patch and whole cell calcium currents recorded simultaneously in snail neurons. *Journal of General Physiology* **83**, 727–50.

Lux, H. D. and Eckert, R. (1974). Inferred slow inward current in snail neurones. *Nature* **250**, 574–6.

Lux, H. D. and Hofmeier, G. (1982*a*). Properties of a calcium- and voltage-activated potassium current in *Helix pomatia* neurons. *Pfluegers Archiv* **394**, 61–9.

Lux, H. D. and Hofmeier, G. (1982*b*). Activation characteristics of the calcium-dependent outward potassium current in *Helix*. *Pfluegers Archiv* **394**, 70–7.

Lux, H. D. and Nagy, K. (1981). Single channel Ca^{2+} currents in *Helix pomatia* neurons. *Pfluegers Archiv* **391**, 252–4.

Lux, H. D., Neher, E., and Marty, A. (1981). Single channel activity associated with the calcium dependent outward current in *Helix pomatia*. *Pfluegers Archiv* **389**, 293–5.

Lynch, G. and Baudry, M. (1984). The biochemistry of memory: a new and specific hypothesis. *Science* **224**, 1057–63.

McAfee, D. A., Henon, B. K., Horn, J. P., and Yarowsky, P. (1981). Calcium currents modulated by adrenergic receptors in sympathetic neurons. *Federation Proceedings* **40**, 2246–9.

McCleskey, E. W. and Almers, W. (1985). The Ca channel in skeletal muscle is a large pore. *Proceedings of the National Academy of Sciences USA* **82**, 7149–53.

McCleskey, E. W. *et al.* (1987). ω-Conotoxin: direct and persistent blockade of specific types of calcium channels in neurons but not muscle. *Proceedings of the National Academy of Sciences USA* **84**, 4327–31.

MacDermott, A. B., Mayer, M. L., Westbrook, G. L., Smith, S. J., and Barker, J. L. (1986). NMDA-receptor activation increases cytoplasmic calcium concentration in cultured spinal cord neurones. *Nature* **321**, 519–22.

Macdonald, R. L. and Werz, M. A. (1986). Dynorphin A decreases voltage-dependent calcium conductance of mouse dorsal root ganglion neurones. *Journal of Physiology (London)* **377**, 237–49.

Macdonald, R. L., Skerritt, J. H., and Werz, M. A. (1986). Adenosine agonists reduce voltage-dependent calcium conductance of mouse sensory neurones in cell culture. *Journal of Physiology (London)* **370**, 75–90.

McFadzean, I. (1988). The ionic mechanisms underlying opioid actions. *Neuropeptides* **11**, 171–81.

McFadzean, I. and Docherty, R. J. (1987). Noradrenaline depresses a high-threshold calcium current in a neuronal cell. *Neuroscience Letters* Suppl. 29, 24.

McLaughin, S. G. A., Mulrine, N., Gresalfi, T., Vaio, G., and McLaughlin, A. (1981). Adsorption of divalent cations to bilayer membranes containing phosphatidylserine. *Journal of General Physiology* **77**, 445–73.

McLaughlin, S. G., Szabo, G., and Eisenman, G. (1971). Divalent ions and the surface potential of charged phospholipid membranes. *Journal of General Physiology* **58**, 667–87.

MacVicar, B. A. and Tse, F. W. Y. (1988). Norepinephrine and cyclic adenosine 3′,5′-cyclic monophosphate enhance a nifedipine-sensitive calcium current in cultured rat astrocytes. *Glia* **1**, 359–65.

Magleby, K. L. and Pallotta, B. S. (1983*a*). Calcium dependence of open and shut interval distributions from calcium-activated potassium channels in cultured rat muscle. *Journal of Physiology (London)* **344**, 585–604.

Magleby, K. L. and Pallotta, B. D. (1983*b*). Burst kinetics of single calcium-activated potassium channels in cultured rat muscle. *Journal of Physiology (London)* **344**, 605–23.

Magura, I. S. (1977). Long-lasting inward currents in snail neurones in barium solutions in voltage-clamp conditions. *Journal of Membrane Biology* **35**, 239–56.

Marchetti, C. and Brown, A. M. (1988). Protein kinase activator 1-oleoyl-2-acetyl-sn-glycerol inhibits two types of calcium currents in GH3 cells. *American Journal of Physiology* **254**, 206–10.

Marchetti, C., Carbone, E., and Lux, H. D. (1986). Effects of dopamine and noradrenaline on Ca channels of cultured sensory and sympathetic neurons of chick. *Pfluegers Archiv* **406**, 104–11.

Maricq, A. V. and Korenbrot, J. I. (1988). Calcium and calcium-dependent chloride currents generate action potentials in solitary cone photoreceptors. *Neuron* **1**, 503–15.

Markin, V. S. and Chizmadjev, Yu. A. (1974). *Induced ion transport*. Publishing House 'Nauka', Moscow.

Marom, S. and Dagan, D. (1987). Calcium current in growth balls from isolated *Helix aspersa* neuronal growth cones. *Pfluegers Archiv* **409**, 578–81.

Marqueze, B., Martin-Mouton, N., Leveque, C., and Couraud, F. (1988). Characterization of the ω-conotoxin-binding molecule in rat brain synaptosomes and cultured neurons. *Molecular Pharmacology* **34**, 87–90.

Martell, A. E. and Smith, R. M. (1977). *Critical stability constants*. Plenum Press, New York.

Marty, A. (1983). Ca^{2+}-dependent K^+ channels with large unitary conductance. *Trends in Neurosciences* **6**, 262–5.

Mason, W. T. and Sikdar, S. K. (1989). Characteristics of voltage-gated Ca^{++} currents in ovine gonadotrophs. *Journal of Physiology (London)* **415**, 367–91.

Matsuda, H. (1986). Sodium conductance in calcium channels of guinea-pig ventricular cells induced by removal of external calcium ions. *Pfluegers Archiv* **407**, 465–75.

Matsuda, Y., Yoshida, S., and Yonezawa, T. (1976). A Ca-dependent regenerative response in rodent dorsal root ganglion cells cultured *in vitro*. *Brain Research* **115**, 334–8.

Matsuda, Y., Yoshida, S., and Yonezawa, T. (1978). Tetrodotoxin sensitivity and Ca component of action potentials of mouse dorsal root ganglion cells cultured *in vitro*. *Brain Research* **154**, 69–82.

Matteson, D. R. and Armstrong, C. M. (1984). Na and Ca channels in a transformed line of anterior pituitary cells. *Journal of General Physiology* **83**, 371–94.

Matteson, D. R. and Armstrong, C. M. (1986). Properties of two types of calcium channels in clonal pituitary cells. *Journal of General Physiology* **87**, 161–82.

Mayer, M. L. (1985). A calcium-activated chloride current generates the after-depolarization of rat sensory neurones in culture. *Journal of Physiology (London)* **364**, 217–39.

Meech, R. W. (1974). The sensitivity of *Helix aspersa* neurones to injected calcium ions. *Journal of Physiology (London)* **237**, 259–77.

Meech, R. W. (1978). Calcium-dependent potassium activation in nervous tissues. *Annual Review of Biophysics and Bioengineering* **7**, 1–8.

Meech, R. W. and Standen, N. B. (1975). Potassium activation in *Helix aspersa* neurones under voltage clamp: A component mediated by calcium influx. *Journal of Physiology (London)* **249**, 211–39.

Meldolesi, J., Volpe, P., and Pozzan, T. (1988). The intracellular distribution of calcium. *Trends in Neurosciences* **11**, 449–52.

Mentrard, D., Vassort, G., and Fischmeister, R. (1984). Calcium-mediated inactivation of the calcium conductance in cesium-loaded frog heart cells. *Journal of General Physiology* **83**, 105–31.

Meves, H. (1968). The ionic requirements for the production of action potentials in *Helix pomatia* neurones. *Pfluegers Archiv* **304**, 215–41.

Meves, H. (1975). Calcium currents in squid giant axon. *Philosophical Transactions of the Royal Society of London* **B270**, 377–87.

Meves, H. (1978). Inactivation of the sodium permeability in squid giant nerve fibres. *Progress in Biophysics and Molecular Biology* **33**, 207–30.

Meyer, T., Holowka, D., and Stryer, L. (1988). Highly cooperative opening of calcium channels by inositol 1,4,5-triphosphate. *Science* **240**, 653–6.

Meyers, D. E. R. and Barker, J. L. (1989). Whole-cell patch-clamp analysis of voltage-dependent calcium conductances in cultured embryonic rat hippocampal neurons. *Journal of Neurophysiology* **61**, 467–77.

Mikami, A., *et al.* (1989). Primary structure and functional expression of the cardiac dihydropyridine-sensitive calcium channel. *Nature* **340**, 230–3.

Miledi, R., Parker, I., and Zhu, P. H. (1982). Calcium transients evoked by action potentials in frog twitch muscle fibres. *Journal of Physiology (London)* **333**, 655–80.

Mironov, S. L. (1983*a*). Theoretical model of slow-wave membrane potential oscillations in molluscan neurons. *Neuroscience* **10**, 899–905.

Mironov, S. L. (1983*b*). The effect of the dielectric constant of proteins on the ionic transport processes in biomembranes. *Bioelectrochemistry and Bioenergetics* **10**, 345–56.

Mironov, S. L. and Tepikin, A. V. (1988). On the nature of cyclic changes in the intracellular concentration of Ca^{2+} ions. *Biological Membranes (Moscow)* **5**, 528–35.

Mironov, S. L., Tepikin, A. V., and Grishchenko, A. V. (1985). Two calcium currents in the somatic membrane of the mollusc neurons. *Neurophysiology (Kiev)* **17**, 627–32.

Mitra, R. and Morad, M. (1986). Two types of calcium channels in guinea pig ventricular myocytes. *Proceedings of the National Academy of Sciences USA* **83**, 5340–4.

Miyamoto, T., Ohizumi, Y., Washio, H., and Yasumoto, Y. (1984). Potent excitatory effect of maitotoxin on Ca channels in the insect skeletal muscle. *Pfluegers Archiv* **400**, 439–41.

Miyazaki, S. I., Takahashi, K., and Tsuda, K. (1974). Electrical excitability in the egg cell membrane of the tunicate. *Journal of Physiology (London)* **238**, 37–54.

Miyazaki, S. I., Ohmori, H., and Sasaki, S. (1975). Action potential and non-linear current–voltage relation in starfish oocytes. *Journal of Physiology (London)* **246**, 37–54.

Mo, N., Ammari, R., and Dun, H. J. (1985). Prostaglandin E1 inhibits calcium-dependent potentials in mammalian sympathetic neurons. *Brain Research* **334**, 325–9.

Moczydlowski, E. and Latorre, R. (1983). Gating kinetics of Ca^{2+}-activated K^+ channels from rat muscle incorporated into planar bilayers: Evidence for two voltage-dependent Ca^{2+} binding reactions. *Journal of General Physiology* **82**, 511–42.

Moolenaar, W. H. and Spector, I. (1978). Ionic currents in cultured mouse neuroblastoma cells under voltage-clamp conditions. *Journal of Physiology (London)* **278**, 265–86.

Morad, M. and Lux, H. D. (1987). Single unit analysis of the proton-induced transformation of Ca^{2+} channel in sensory neuron. *Biophysical Journal* **51**, (2), 32.

Morad, M., Davies, N. M., Kaplan, J. H., and Lux, H. D. (1988). Inactivation and block of calcium channels by photo-released Ca^{2+} in dorsal root ganglion neurons. *Science* **241**, 842–4.

Moriarty, C. M. (1978). Role of calcium in the regulation of adenohypophyseal hormone release. *Life Sciences* **23**, 185–94.

Moskowitz, N., Glassman, A., Ores, C., Schook, W., and Puszkin, S. (1983). Phosphorylation of brain synaptic and coated vesicle proteins by endogenous Ca^{2+}/calmodulin- and cAMP-dependent protein kinases. *Journal of Neurochemistry* **40**, 711–18.

Mozhayeva, G. N. and Naumov, A. P. (1972*a*). The effect of surface charge on stationary potassium conductance of the Ranvier node membrane. I. Changes in the extracellular pH. *Biophysics (Moscow)* **17**, 412–20.

Mozhayeva, G. N. and Naumov, A. P. (1972*b*). The effect of surface charge on stationary potassium conductance of the Ranvier node membrane. II. Changes in ionic strength of the extracellular solution. *Biophysics (Moscow)* **17**, 618–22.

Mullins, L. J. (1977). A mechanism for Na/Ca transport. *Journal of General Physiology* **70**, 681–95.

Mullins, L. J. and Requena, J. (1979). Calcium measurement in the periphery of an axon. *Journal of General Physiology* **74**, 393–413.

Murphy, K. M., Gould, K. J., and Snyder, S. H. (1982). Autoradiographic visualization of (^{3}H)-nitrendipine binding sites in rat brain: Localization to synaptic zones. *European Journal of Pharmacology* **81**, 517–19.

Murphy, K. M., Gould, R. J., Larcent, B. L., and Snyder, S. H. (1983). A unitary mechanism of calcium antagonist drug action. *Proceedings of the National Academy of Sciences USA* **80**, 860–4.

Nachshen, D. A. (1984). Selectivity of the Ca binding site in synaptosome Ca channels. *Journal of General Physiology* **83**, 941–67.

Nachshen, D. A. and Blaustein, M. P. (1980). Some properties of potassium-stimulated calcium influx in presynaptic nerve endings. *Journal of General Physiology* **76**, 709–28.

Nachshen, D. A. and Blaustein, M. P. (1982). Influx of calcium, strontium and barium in presynaptic nerve endings. *Journal of General Physiology* **79**, 1065–87.

Narahashi, T., Tsunoo, A., and Yoshii, M. (1987). Characterization of two types of calcium channels in mouse neuroblastoma cells. *Journal of Physiology (London)* **383**, 231–49.

Naruševicius, E. V. and Rapoport, M. S. (1982). Energetic profile of the open calcium channel: Experimental and theoretical studies. *General Physiology and Biophysics* **1**, 103–15.

Neale, J. H. and Barker, J. L. (1977). Bidirectional axonal transport of $^{45}Ca^{2+}$: studies in isolated frog sensory motor and sympathetic neurons, *Aplysia* cerebral ganglion and the goldfish visual system. *Brain Research* **129**, 45–59.

Nedergaard, S., Bolam, J. P., and Greenfield, S. A. (1988). Facilitation of a dendritic calcium conductance by 5-hydroxytryptamine in the substantia nigra. *Nature* **333**, 174–7.

Nedergaard, S., Webb, C., and Greenfield, S. A. (1989). A possible ionic basis for dendritic release of dopamine in the guinea-pig substantia nigra. *Acta Physiologica Scandinavica* **135**, 67–8.

Nelson, P., Ruffner, W., and Nirenberg, M. (1969). Neuronal tumor cells with excitable membranes grown *in vitro*. *Proceedings of the National Academy of Sciences USA* **64**, 1004.

Niedergerke, R. and Orkand, R. K. (1966). The dependence on the action potential of the frog's heart on the external and intracellular sodium concentration. *Journal of Physiology (London)* **184**, 312–34.

Nishi, K., Akaike, N., Oyama, Y., and Ito, H. (1983). Actions of calcium antagonists on calcium currents in *Helix* neurons: Specificity and potency. *Circulation Research* **52**, (2), 1-53–1-59.

Nowycky, M. C., Fox, A. P., and Tsien, R. W. (1985*a*). Three types of neuronal calcium channel with different calcium agonist sensitivity. *Nature* **316**, 440–3.

Nowycky, M. C., Fox, A. P., and Tsien, R. W. (1985*b*). Long-opening mode of gating of neuronal calcium channels and its promotion by the dihydropyridine calcium agonist Bay K 8644. *Proceedings of the National Academy of Sciences USA* **82**, 2178–82.

Obaid, A. L., Flores, R., and Salzberg, B. M. (1989). Calcium channels that are required for secretion from intact nerve terminals of vertebrates are sensitive to ω-conotoxin and relatively insensitive to dihydropyridines. Optical studies with and without voltage-sensitive dyes. *Journal of General Physiology* **93**, 715–29.

O'Callahan, C. M. and Hosey, M. M. (1988). Multiple phosphorylation sites in the 165-kilodalton peptide associated with dihydropyridine-sensitive calcium channels. *Biochemistry* **27**, 6071–7.

Ochs, S. (1982). Calcium and the mechanism of axoplasmic transport. *Federation Proceedings* **41**, 2301–6.

Ochs, S., Chan, S. Y., Worth, R. M., and Jersild, R. (1978). Calcium and fast axoplasmic transport in mammalian nerve. *Biophysical Journal* **21**, 187a.

Ogura, A., Miyamoto, M., and Kudo, Y. (1988). Neuronal death *in vitro*: parallelism between survivability of hippocampal neurones and sustained elevation of cytosolic Ca^{++} after exposure to glutamate receptor agonist. *Experimental Brain Research* **73**, 447–59.

Ohki, S. (1978). Ionic conductances and surface potential of axon membranes. *Biophysical Journal* **21**, 163a.

Ohki, S. and Kurland, R. (1981). Surface potential of phosphatidylserine monolayers. II. Divalent and monovalent ion binding. *Biochimica et Biophysica Acta* **645**, 170–6.

Ohmori, H. (1984). Studies of ionic currents in the isolated vestibular hair cells of the chick. *Journal of Physiology (London)* **350**, 561–81.

Ohmori, H. and Yoshii, M. (1977). Surface potential reflected in both gating and permeation mechanisms of sodium and calcium channels of the tunicate egg cell membrane. *Journal of Physiology (London)* **267**, 429–63.

Ohnishi, S. T. (1978). Problems in determinating ionized calcium concentration by arsenazo III. *Biophysical Journal* **21**, 186a.

Ohya, Y., Kitamura, K., and Kuriyama, H. (1987). Modulation of ionic currents in smooth muscle bands of the rabbit intestine by intracellularly perfused ATP and cyclic AMP. *Pfluegers Archiv* **408**, 465–73.

Okamoto, H., Takahashi, K., and Yoshii, M. (1976*a*). Membrane currents of the tunicate egg under the voltage-clamp condition. *Journal of Physiology (London)* **254**, 607–38.

Okamoto, H., Takahashi, K., and Yoshii, M. (1976*b*). Two components of the calcium current in the egg cell membrane of the tunicate. *Journal of Physiology (London)* **255**, 527–61.

Okamoto, H., Takahashi, K., and Yamashita, N. (1977). Ionic currents through the membrane of the mammalian oocyte and their comparison with those in the tunicate and sea urchin. *Journal of Physiology (London)* **267**, 465–95.

Okamoto, K., Kimura, H., and Sakai, Y. (1983). Effects of taurine and GABA on Ca spikes and Na spikes in cerebellar Purkinje cells *in vitro*: Intrasomatic study. *Brain Research* **260**, 249–59.

Olivera, B. M. *et al.* (1987). Neuronal calcium channel antagonists. Discrimination between calcium channel subtypes using ω-conotoxin from *Conus magus* venom. *Biochemistry* **26**, 2086–90.

Oomura, Y., Ozaki, S., and Maeno, T. (1961). Electrical activity of a giant cell under abnormal conditions. *Nature* **191**, 1265–7.

Osborne, N. N., Tobin, A. B., and Ghazi, H. (1988). Role of inositol triphosphate as a second messenger in signal transduction processes: an essay. *Neurochemical Research* **13**, 177–91.

Ottoson, D. and Swerup, C. (1982). Studies on the role of calcium in adaptation of the crustacean stretch receptor: Effects of intracellular injection of calcium, EGTA and EDTA. *Brain Research* **244**, 337–41.

Owen, D. G., Segal, M., and Barker, J. L. (1984). A Ca-dependent Cl^- conductance in cultured mouse spinal neurones. *Nature* **311**, 567–70.

Oyama, Y., Akaike, N., and Nishi, K. (1986). Effects of *n*-alkanols on the calcium current of intracellularly perfused neurons of *Helix aspersa*. *Brain Research* **376**, 280–4.

Ozawa, S., Tsuzuki, K., Iino, M., Ogura, A., and Kudo, Y. (1989). Three types of voltage-dependent calcium current in cultured rat hippocampal neurons. *Brain Research* **495**, 329–36.

Palade, P. T. and Almers, W. (1978). Slow Na^+ and Ca^{2+} currents across the membrane of frog skeletal muscle fibres. *Biophysical Journal* **21**, 168a.

Panfili, E., *et al.* (1976). Specific inhibition of mitochondrial Ca^{2+} transport by antibodies directed to the Ca^{2+}-binding glycoprotein. *Nature* **264**, 185–6.

Paupardin-Tritsch, D., Hammond, C., and Gerschenfeld, H. M. (1986*a*). Serotonin and cyclic GMP both induce an increase of the calcium current in the same identified molluscan neurons. *Journal of Neuroscience* **6**, 2715–23.

Paupardin-Tritsch, D., Hammond, C., Gerschenfeld, H. M., Nairn, A. C., and Greengard, P. (1986*b*). cGMP-dependent protein kinase enhances Ca^{2+} current and potentiates the serotonin-induced Ca^{2+} current increase in snail neurones. *Nature* **323**, 812–14.

Pellmar, T. C. (1981). Ionic mechanism of a voltage-dependent current elicited by cyclic AMP. *Cellular and Molecular Neurobiology* **1**, 87–97.

Pellmar, T. C. and Carpenter, D. O. (1979). Voltage-dependent calcium current induced by serotonin. *Nature* **277**, 483–4.

Pelzer, D., Trautwein, W., and McDonald, T. F. (1982). Calcium channel block and recovery from block in mammalian ventricular muscle treated with organic channel inhibitors. *Pflügers Archiv* **394**, 97–105.

Peres, A., Sturani, E., and Zippel, R. (1988). Properties of the voltage-dependent calcium channel of mouse Swiss 3T3 fibroblasts. *Journal of Physiology (London)* **401**, 639–55.

Perez-Reyes, E., *et al.* (1989). Induction of calcium currents by the expression of the α_1-subunit of the dihydropyridine receptor from skeletal muscle. *Nature* **340**, 233–6.

Peters, K. A., Demaille, J. G., and Fischer, E. H. (1977). Adenosine 3′,5′-monophosphate dependent protein kinase from bovine heart: Characterization of the catalytic subunit. *Biochemistry* **16**, 5691–7.

Peyer, J. E. de, Cachelin, A. B., Levitan, I. B., and Reuter, H. (1982). Ca^{2+}-activated K^+ conductance in internally perfused snail neurons is enhanced by protein phosphorylation. *Proceedings of the National Academy of Sciences USA* **79**, 4207–11.

Philipson, K. D. (1983). "Calciductin" and voltage-sensitive calcium uptake. *Journal of Molecular and Cellular Cardiology* **15**, 867–9.

Pitman, R. M. (1975). Calcium dependent action potentials in the cell body of an insect motoneurone. *Journal of Physiology (London)* **251**, 62P–63P.

Plant, T. D. and Standen, N. B. (1981). Calcium-current inactivation in identified neurones of *Helix aspersa*. *Journal of Physiology (London)* **321**, 273–85.

Plant, T. D., Standen, N. B., and Ward, T. A. (1983). Calcium injection and calcium channel inactivation. In *Neurology and neurobiology*, Vol. 5, *The physiology of excitable cells* (ed. A. D. Grinnell and W. J. Moody, Jr), pp. 39–49. Alan R. Liss, New York.

Pogorelaya, N. Ch. and Fomina, A. F. (1984). The effect of pH changes in the medium on differentiation of cultured neuroblastoma cells. *Neurophysiology (Kiev)* **16**, 842–5.

Pogorelaya, N. Ch., Skibo, G. G., and Troitzkaya, N. K. (1980). Structural peculiarities of isolated and perfused neurons of *Helix pomatia* mollusc. *Neurophysiology (Kiev)* **12**, 297–302.

Ponomarev, V. N., Narusevicius, E. V., and Chemeris, N. K. (1980). Blocking action of Ni, Co, Mn and Mg ions on inward current through calcium channels in *Limnaea* neurons. *Neurophysiology (Kiev)* **12**, 221–3.

Pott, L. and Lipp, P. (1987). Dual effect of the Ca^{2+} calmodulin antagonist fendiline on Ca current in single cardiac cells. *Journal of Molecular and Cellular Cardiology* **19**, (3), 75.

Pow, D. V. and Morris, J. F. (1989). Dendrites of hypothalamic magnocellular neurons release neurohypophysial peptides by exocytosis. *Neuroscience* **32**, 435–9.

Powell, T. and Twist, V. W. (1976). A rapid technique for the isolation and purification of adult cardiac muscle cells having respiratory control and a tolerance to calcium. *Biochemical and Biophysical Research Communications* **72**, 327–33.

Puszkin, S., Niklas, W. J., and Berl, S. (1972). Actomyosin-like protein in brain: Subcellular distribution. *Journal of Neurochemistry* **19**, 1319–33.

Rajs, J. *et al.* (1978). A rapid method for the isolation of viable cardiac myocytes from adult rat heart. *Experimental Cell Research* **115**, 183–9.

Rane, S. G., Holz, G. G., and Dunlap, K. (1987). Dihydropyridine inhibition of neuronal calcium current and substance P release. *Pfluegers Archiv* **409**, 361–6.

Ransom, B. R. and Holz, R. W. (1977). Ionic determinants of excitability in cutured mouse dorsal root ganglion and spinal cord cells. *Brain Research* **136**, 445–53.

Rasmussen, H., Clayberger, C., and Gustin, M. C. (1979). The messenger function of calcium in cell activation. In *Secretory Mechanisms*, Symposia of the Society for Experimental Biology, No. XXXIII, pp. 161–97. Cambridge University Press.

Rasmussen, H., Kojima, I., Kojima, K., Zawalich, W., and Apfeldorf, W. (1985). Calcium as intracellular messenger; sensitivity modulation, C-kinase pathway, and sustained cellular response. In *Advances in cyclic nucleotide and protein phosphorylation research*, Vol. 18 (ed. P. Greengard and G. A. Robinson), pp. 159–93. Raven, New York.

Requena, J. and Mullins, L. J. (1979). Calcium movement in nerve fibres. *Quarterly Review of Biophysics* **12**, 371–460.

Requena, J., DiPolo, R., Brinley, F. J., and Mullins, L. J. (1977). The control of ionized calcium in squid axons. *Journal of General Physiology* **70**, 329–53.

Reuter, H. (1965). Ueber der Wirkung von Adrenalin auf den cellularen Ca-Umsatz des Meerschweinchenvorhofs. *Naunyn-Schmiederbergs Archives of Pharmacology* **251**, 401–12.

Reuter, H. (1967). The dependence of slow inward current in Purkinje fibres on the extracellular calcium concentration. *Journal of Physiology (London)* **192**, 479–92.

Reuter, H. (1974). Localization of bete adrenergic receptors and effects of noradrenaline and cyclic nucleotides on action potentials, ionic currents and tension in mammalian cardiac muscle. *Journal of Physiology (London)* **242**, 429–51.

Reuter, H. (1979). Properties of two inward membrane currents in the heart. *Annual Review of Physiology* **41**, 413–24.

Reuter, H. (1987). Modulation of ion channels by phosphorylation and second messengers. *News in Physiological Sciences* **2**, 168–71.

Reuter, H. and Scholz, H. (1977*a*). A study of the ion selectivity and the kinetic properties of the calcium dependent slow inward current in mammalian cardiac muscle. *Journal of Physiology (London)* **264**, 17–47.

Reuter, H. and Scholz, H. (1977*b*). The regulation of the calcium conductance of cardiac muscle by adrenaline. *Journal of Physiology (London)* **264**, 49–62.

Reuter, H. and Seitz, N. (1968). The dependence of calcium efflux from cardiac muscle on temperature and external ion composition. *Journal of Physiology (London)* **195**, 451–70.

Reuter, H., Stevens, C. F., Tsien, R. W., and Yellen, G. (1982). Properties of single calcium channels in cardiac cell culture. *Nature* **297**, 501–4.

Reymann, K. G., Matthies, H. K., Schulzeck, K., and Matthies, H. (1989). *N*-Methyl-D-aspartate receptor activation is required for the induction of both early and late phases of long-term potentiation in rat hippocampal slices. *Neuroscience Letters* **96**, 96–101.

Reynolds, J. N. and Carlen, P. L. (1989). Diminished calcium currents in aged hippocampal dentate gyrus granule neurons. *Brain Research* **479**, 384–90.

Rinaldi, M. L., LePeuch, C. J., and Demaille, J. G. (1981). The epinephrine-induced activation of the cardiac slow Ca^{2+} channel is mediated by the cAMP-dependent phosphorylation of calciductin, a 23 000 Mr sarcolemmal protein. *FEBS Letters* **129**, 277–81.

Ringer, S. (1883). A further contribution regarding the influence of the different constituents of the blood on the contraction of the heart. *Journal of Physiology (London)* **4**, 29–42.

Rodriguez, R., *et al.* (1988). Histamine-stimulated synaptosomal Ca^{2+} uptake through activation of calcium channels. *Biochemical and Biophysical Research Communications* **153**, 1136–43.

Rogers, J. *et al.* (1983). Intracellular pH and free calcium changes in single cells using quin 1 and quin 2 probes and fluorescence microscopy. *FEBS Letters* **161**, 21–7.

Rohrkasten, A., Meyer, H., Nastainczuk, W., Sieber, M., and Hofmann, F. (1988). cAMP-dependent protein kinase rapidly phosphorylates serine-687 of the skeletal muscle receptor for calcium channel blockers. *Journal of Biological Chemistry* **263**, 15325–9.

Romey, G., Garcia, L., Dimitriadou, V., Pincon-Raymond, M., Rieger, F., and Lazdunski, M. (1989). Ontogenesis and localization of Ca^{2+} channels in mammalian skeletal muscle in culture and role in excitation–contraction coupling. *Proceedings of the National Academy of Sciences USA* **86**, 2933–7.

Rorsman, P. (1988). Two types of Ca^{2+} currents with different sensitivities to organic

Ca^{2+} channel antagonists in guinea pig pancreatic α_2 cells. *Journal of General Physiology* **91**, 243–54.

Rosenthal, W., Hescheler, J., Hinsch, K. D., and Schultz, G. (1989). Bidirectional hormonal modulation of voltage-dependent Ca^{++} channels. *Drug Research* **39-1**, 143–8.

Ross, W. N. and Werman, R. (1987). Mapping calcium transients in the dendrites of Purkinje cells from the guinea-pig cerebellum *in vitro*. *Journal of Physiology (London)* **389**, 319–36.

Rubtsov, A. M. and Murphy, A. J. (1988). Caffeine interaction with the Ca-release channels of heavy sarcoplasmic reticulum. Evidence that 170 kD Ca-binding protein is a caffeine receptor of the Ca-channels. *Biochemical and Biophysical Research Communications* **151**, 462–8.

Ruth, P., *et al.* (1989). Primary structure of the β subunit of the DHP-sensitive calcium channel from skeletal muscle. *Science* **245**, 1115–18.

Saida, K. and van Breemen, C. (1985). Cyclic nucleotides and calcium movements. In *Calcium entry blockers and tissue protection* (ed. T. Godfraind, P. M. Vanhoutte, S. Govoni, and R. Paoletti), pp. 41–50. Raven, New York.

Saida, K. and van Breeman, C. (1987). GTP requirement for inositol-1,4,5-triphosphate-induced Ca^{2+} release from sarcoplasmic reticulum in smooth muscle. *Biochemical and Biophysical Research Communications* **144**, 1313–16.

Sanchez, J. A. and Stefani, E. (1978). Inward calcium current in twitch muscle fibres of the frog. *Journal of Physiology (London)* **283**, 197–209.

Sanchez, J. A. and Stefani, E. (1983). Kinetic properties of calcium channels of twitch muscle fibres of the frog. *Journal of Physiology (London)* **337**, 1–17.

Sand, O., Sletholt, K., Gautnik, K., and Hang, E. (1983). Trifluoperazine blocks calcium-dependent action potentials and inhibits hormone release from rat pituitary tumour cells. *European Journal of Pharmacology* **86**, 177–84.

Sasakawa, N., Nakaki, T., Yamomoto, S., and Kato, R. (1989). Stimulation by ATP of inositol triphosphate accumulation and calcium mobilization in cultured adrenal chromaffin cells. *Journal of Neurochemistry* **52**, 441–7.

Schneider, A. S., Cline, H. T., Rosenheck, K., and Sonenberg, M. (1981). Stimulus-secretion coupling in isolated adrenal chromaffin cells: calcium channel activation and possible role of cytoskeletal elements. *Journal of Neurochemistry* **37**, 567–75.

Schramm, M., Thomas, G., Towart, R., and Franckowiak, G. (1983). Novel dihydropyridines with positive inotropic action through activation of Ca^{2+} channels. *Nature* **303**, 535–7.

Schwartzkroin, P. A. and Slawsky, M. (1977). Probable calcium spikes in hippocampal neurons. *Brain Research* **135**, 157–61.

Schwindt, P. C. and Crill, W. E. (1980). Properties of a persistent inward current in normal and TEA-injected montoneurons. *Journal of Neurophysiology* **43**, 1700–24.

Scott, B. S. and Edwards, B. A. V. (1980). Electric membrane properties of adult mouse DRG neurons and the effect of culture duration. *Journal of Neurobiology* **11**, 291–301.

Scott, R. H. and Dolphin, A. C. (1986). Regulation of calcium currents by GTP analogue: potentiation of (−)-baclofen-mediated inhibition. *Neuroscience Letters* **69**, 59–64.

Scott, R. H. and Dolphin, A. C. (1987). Activation of a G protein promotes agonist responses to calcium channel ligands. *Nature* **330**, 760–2.

Seaman, G. V. F., Vassar, P. S., and Kendall, M. J. (1969). Electrophoretic studies of human polynuclear leucocytes and erythrocytes. *Archives of Biochemistry and Biophysics* **135**, 356–63.

Semin, B. K. and Ivanov, I. I. (1989). Inhibition of the voltage-dependent calcium channel by calmodulin antagonists: amino acid sequence of receptor for calcium channel blockers has the segment similar to antagonist binding region in calmodulin. *Biological Membranes (Moscow)* **6**, 892–3.

Severin, E. S., *et al*. (1981). The mechanism of action of brain cyclic AMP-dependent protein kinase: Structure of active sites, mechanism of dissociation of the holoenzyme, and possible biological role of its subunits. In *Proceedings in Life Science* (ed. H. Holzerp), pp. 44–58. Springer, Berlin.

Shakhovalov, Yu. A. (1977). Quantitative description of the Na and Ca inward currents in molluscan neurones by Hodgkin–Huxley equations. *Journal of Physiology (London)* **270**, 562–7.

Shiina, T., Wayne, R., Lim Tang, H. Y., and Tazawa, M. (1988). Possible involvement of protein phosphorylation/dephosphorylation in the modulation of Ca^{2+} channel in tonoplast-free cells of *Nitellopsis*. *Journal of Membrane Biology* **102**, 255–64.

Shimahara, T. and Icard-Liepkalns, C. (1987). Activation of enkaphalin receptors reduces calcium conductance in neuroblastoma cells. *Brain Research* **415**, 357–61.

Shimahara, T. and Tauc, L. (1977). Cyclic AMP induced by serotonin modulates the activity of an identified synapse in *Aplysia* by facilitating the active permeability to calcium. *Brain Research* **127**, 168–72.

Shimomura, O., Johnson, F. H., and Saiga, Y. (1962). Extraction, purification and properties of aequorin, a bioluminescent protein from the luminous hydromedusan, *Auquorea*. *Journal of Cellular and Comparative Physiology* **59**, 233–9.

Shuba, Ya. M. and Lyubanova, O. P. (1988). Calcium permeability of PC12 pheochromocytoma cells. *Biological Membranes (Moscow)* **5**, 698–710.

Shuba, Ya. M. and Savchenko, A. N. (1987). Two conductance states of single calcium channels in the membranes of dorsal root ganglion cells from mouse embryos. *Biological Membranes (Moscow)* **4**, 374–8.

Shuba, Ya. M. and Teslenko, V. I. (1987). Kinetic model for activation of single calcium channels in mammalian sensory neurone membrane. *Biological Membranes (Moscow)* **4**, 315–29.

Shuba, Ya. M., Hesslinger, B., Trautwein, W., McDonald, T. F., and Pelzer, D. (1990). Whole-cell calcium current in guinea-pig ventricular myocytes dialysed with guanine nucleotides. *Journal of Physiology (London)* **424**, 205–228.

Siegelbaum, S. A. and Tsien, R. W. (1980). Calcium-activated transient outward current in calf cardiac Purkinje fibres. *Journal of Physiology (London)* **299**, 485–506.

Sigworth, F. J. (1977). Na-current fluctuations give estimates of single-channel conductance in frog nerve. *Biophysical Journal* **17**, 10a.

Simashko, S. M., Weiland, G. A., and Oswald, R. E. (1988). Pharmacological characterization of two calcium currents in GH3 cells. *American Journal of Physiology* **254**, E328–E336.

Smith, A. D. (1971). Summing up: some implications of the neuron as a secreting cell. *Philosophical Transactions of the Royal Society B*, **261**, 423–37.

Smith, J. S., Imagawa, T., Ma, J., Fill, M., Campbell, K. P., and Coranddo, R. (1988).

Purified ryanodine receptor from rabbit skeletal muscle is the calcium-release channel of sarcoplasmic reticulum. *Journal of General Physiology* **92**, 1–26.

Song, S.-Y., Saito, K., Noguchi, K., and Konishi, S. (1989). Different GTP-binding proteins mediate regulation of calcium channels by acetylcholine and noradrenaline in rat sympathetic neurons. *Brain Research* **494**, 383–6.

Spector, I. (1981). Electrophysiology of clonal nerve cell lines. In *Excitable cells in tissue culture*, (ed. P. G. Nelson and S. Lieberman), pp. 247–77. Plenum Press, New York.

Spitzer, N. C. and Lamborghini, J. E. (1976). The development of the action potential mechanism of amphibian neurons isolated in culture. *Proceedings of the National Academy of Sciences USA* **73**, 1641–5.

Standen, N. B. (1975*a*). Calcium and sodium ions as charge carriers in the action potential of an identified snail neurone. *Journal of Physiology (London)* **249**, 241–52.

Standen, N. B. (1975*b*). Voltage-clamp studies of the calcium inward current in an identified snail neurone: Comparison with the sodium-inward current. *Journal of Physiology (London)* **249**, 253–68.

Standen, N. B. (1981). Ca channel inactivation by intracellular Ca injection into *Helix* neurones. *Nature* **293**, 158–9.

Standen, N. B. and Stanfield, P. R. (1982). A binding-site model for calcium channel inactivation that depends on calcium entry. *Proceedings of the Royal Society of London* **B217**, 101–10.

Stern, O. (1924). Theory of a double-electric layer with the consideration of the adsorption processes. *Zeitschrift für Elektrochemie* **30**, 508–16.

Stevens, C. F. (1972). Inferences about membrane properties from electrical noise measurements. *Biophysical Journal* **12**, 1028–47.

Streit, J. and Lux, H. D. (1987). Voltage dependent calcium currents in PC12 growth cones and cells during NGF-induced cell growth. *Pfluegers Archiv* **408**, 634–41.

Strong, J. A., Fox, A. P., Tsien, R. W., and Kaczmarek, I. K. (1987). Stimulation of protein kinase C recruits covert calcium channels in *Aplysia* bag cell neurons. *Nature* **325**, 714–17.

Suarez-Kurtz, G., Katz, G. M., and Reuben, J. P. (1987). Currents carried by sodium ions through transient calcium channels in clonal GH3 pituitary cells. *Pfluegers Archiv* **410**, 345–7.

Sugaya, E. and Onozuka, M. (1978). Unequal distribution of calcium and magnesium of snail neuron. *Experientia* **34**, 1299–301.

Suszkiw, J. B., O'Leary, M. E., Murawsky, M. M., and Wang, T. (1986). Presynaptic calcium channels in rat cortical synaptosomes: fast-kinetics of phasic calcium influx, channel inactivation, and relationship to nitrendipine receptors. *Journal of Neuroscience* **6**, 1349–57.

Suszkiw, J. B., Murawsky, M. M., and Fortner, R. C. (1987). Heterogeneity of presynpatic calcium channels revealed by specific differences in the sensitivity of synaptosomal ^{45}Ca entry to ω-conotoxin. *Biochemical and Biophysical Research Communications* **145**, 1283–6.

Suszkiw, J. B., Murawsky, M. M., and Shi, M. (1989). Further characterization of phasic calcium influx in rat cerebrocortical synaptosomes: interferences regarding calcium channel type(s) in nerve endings. *Journal of Neurochemistry* **52**, 1260–9.

Suzuki, N. and Yoshioka, T. (1987). Differential blocking action of synthetic ω-

conotoxin on components of Ca^{2+} channel current in clonal GH3 cells. *Neuroscience Letters* **75**, 235–9.

Swandulla, D. and Armstrong, C. M. (1988). Fast-deactivating calcium channels in chick sensory neurons. *Journal of General Physiology* **92**, 197–218.

Swandulla, D., Carbone, E., Schaefer, K., and Lux, H. D. (1987). Effect of menthol on two types of Ca currents in cultured sensory neurons of vertebrates. *Pfluegers Archiv* **409**, 52–9.

Takahashi, K. and Yoshii, M. (1978). Effects of internal free calcium upon the sodium and calcium channels in the tunicate egg analysed by the internal perfusion technique. *Journal of Physiology (London)* **279**, 519–49.

Takahashi, K. and Yoshii, M. (1981). Development of sodium, calcium and potassium channels in the cleavage-arrested embryo of an ascidian. *Journal of Physiology (London)* **315**, 515–29.

Takahashi, M., Seagar, M. J., Jones, J. F., Reber, B. F., and Catterall, W. A. (1987). Subunit structure of dihydropyridine-sensitive calcium channels from skeletal muscle. *Proceedings of the National Academy of Sciences USA* **84**, 5478–82.

Tang Cha-Min, Presser, F., and Morad, M. (1988). Amiloride selectively blocks the low threshold (T) calcium channel. *Science* **240**, 213–15.

Taraskevich, P. S. and Douglas, W. W. (1989). Effects of BAY K 8644 on Ca-channel currents and electrical activity in mouse melanotrophs. *Brain Research* **491**, 102–8.

Taylor, S. J., Bartlett, M. J., and Greenfield, S. A. (1988). Release of acetylcholinesterase within the guinea-pig substantia nigra: effects of 5-hydroxytryptamine and amphetamine. *Neuropharmacology* **27**, 507–14.

Tepikin, A. V., Belan, P. V., and Mironov, S. L. (1987). Changes in intracellular Ca^{2+} in isolated snail neurones revealed by fluorescent probe Fura-2. *Biological Membranes (Moscow)* **4**, 882–9.

Thayer, S. A. and Fairhurst, A. S. (1983). The interaction of dihydropyridine calcium channel blockers with calmodulin and calmodulin inhibitors. *Molecular Pharmacology* **24**, 6–9.

Thayer, S. A., Hirning, L. D., and Miller, R. J. (1988). The role of caffeine-sensitive calcium stores in the regulation of the intracellular free calcium concentration in rat sympathetic neurons *in vitro*. *Molecular Pharmacology* **34**, 664–73.

Thomas, M. V. (1984). Voltage-clamp analysis of a calcium-mediated potassium conductance in cockroach (*Periplaneta americana*) central neurones. *Journal of Physiology (London)* **350**, 158–78.

Tillotson, D. (1979). Inactivation of Ca-conductance dependent on entry of Ca-ions in molluscan neurones. *Proceedings of the National Academy of Sciences USA* **76**, 1497–500.

Tillotson, D. and Gorman, A. L. F. (1980). Non-uniform Ca^{2+} buffer distribution in a nerve cell body. *Nature* **286**, 816–17.

Tillotson, D. and Horn, R. (1978). Cs substitution for internal K reveals an inactivating Ca current in *Aplysia* neurons. *Biophysical Journal* **21**, 178a.

Toll, L. (1982). Calcium antagonists: High-affinity binding and inhibition of calcium transport in a clonal cell line. *Journal of Biological Chemistry* **257**, 13189–92.

Toselli, M. and Lux, H. D. (1989). GTP binding proteins mediate acetylcholine inhibition of voltage dependent calcium channels in hippocampal neurons. *Pfluegers Archiv* **413**, 319–21.

Traub, R. D. and Llinas, R. (1979). Hippocampal pyramidal cells: Significance of dendritic ionic conductances for neuronal function and epileptogenesis. *Journal of Neurophysiology* **42**, 476–96.

Trautwein, W., McDonald, T. F., and Tripathi, O. (1975). Calcium conductance and tension in mammalian ventricular muscle. *Pfluegers Archiv* **354**, 55–74.

Trautwein, W., Pelzer, D., and McDonald, T. F. (1983). Interval- and voltage-dependent effects of the calcium channelblocking agents D600 and AQA 39 on mammalian ventricular muscle. *Circulation Research* **52**, 1-60–1-68.

Triggle, D. J. and Janis, R. A. (1984). Nitrendipine: Binding sites and mechanism of action. In *Nitrendipine* (ed. A. Scriabine *et al.*), pp. 33–57. Urban and Schwarzenberg, Baltimore.

Triggle, D. J., Langs, D. A., and Janis, R. A. (1989). Ca^{2+} channel ligands: structure—function relationship of the 1,4-dihydropyridines. *Medicinal Research Reviews* **9**, 123–80.

Tsien, R. W., Giles, W., and Greengard, P. (1972). Cyclic AMP mediates the effects of adrenaline on cardiac Purkinje fibres. *Nature New Biology* **240**, 181–3.

Tsien, R. W., Lipscombe, D., Madison, D. V., Bley, K. R., and Fox, A. P. (1988). Multiple types of neuronal calcium channels and their selective modulation. *Trends Neurosci.* **11**, 431–8.

Tsunoo, A., Yoshii, M., and Narahashi, T. (1984). Two types of calcium channels in neuroblastoma cells and their sensitivities to cyclic AMP. *Society for Neuroscience Abstracts* **10**, (1), 527.

Tsunoo, A., Yoshii, M., and Narahashi, T. (1986). Block of calcium channels by enkaphalin and somatostatin in neuroblastoma-glioma hybrid NG108-15 cells. *Proceedings of the National Academy of Sciences USA* **83**, 9832–6.

Ui, M., Katada, T., Murayama, T., and Kurose, H. (1984). Chemical modification of nucleotide regulatory proteins catalyzed by bacterial toxins. *IUPHAR 9th International Congress of Pharmacology, Abstracts*, pp. S7–3. Macmillan, London.

Undrovinas, A. I. (1984). Perfusion of single cardiomyocytes. In *Intracellular perfusion of excitable cells*, (ed. P. G. Kostyuk and O. A. Krishtal), pp. 53–75. Wiley, Chichester.

Usachev, Yu. M. and Mironov, S. C. (1989). Effect of Sr and Ba ions on Ca binding and transport in nerve cells. *Neurophysiology (Kiev)* **21**, 820–5.

Valeyev, A. E. (1979). Selectivity of calcium channels of the somatic membrane of *Helix pomatia* neurons for calcium, strontium and barium ions. *Neurophysiology (Kiev)* **1**, 371–4.

Vassort, G. (1975). Voltage-clamp analysis of transmembrane ionic currents in guinea-pig myometrium: Evidence for an initial potassium activation triggered by calcium influx. *Journal of Physiology (London)* **252**, 713–34.

Vater, W., *et al.* (1972). Zur Pharmakologie von 4-(2-Nitrophenyl)-2,6-dimethyl-1,4-dihydropyridine-3,5-dicarbonsauredimethylester (Nifedipine). *Arzneimittel-Forschung* **22**, 1–14.

Veselovsky, N. S. and Fedulova, S. A. (1983). Two types of calcium channels in the somatic membrane of rat dorsal root ganglion neurones. *Doklady Akademii Nauk SSSR* **268**, 747–50.

Veselovsky, N. S. and Fomina, A. F. (1986). The study of sodium and calcium channels in the somatic membrane of neuroblastoma cells during their induced differentiation. *Neurophysiology (Kiev)* **18**, 207–14.

Veselovsky, N. S., Kostyuk, P. G., Krishtal, O. A., and Pidoplichko, V. I. (1977).

Separation of ionic currents responsible for generation of action potential in isolated neurons of frog spinal ganglia. *Neurophysiology (Kiev)* **9**, 638–40.

Veselovsky, N. S., Pogorelaya, N.Ch., and Fomina, A. F. (1984). Calcium currents of differentiated neuroblastoma cells. *Neurophysiology (Kiev)* **16**, 527–31.

Veselovsky, N. S., Kostyuk, P. G., Fedulova, S. A,. and Shirokov, R. E. (1985). Deactivation of calcium currents in the somatic membrane of dorsal root ganglion neurons with removal of the membrane potential depolarization shift. *Neurophysiology (Kiev)* **17**, 682–91.

Veselovsky, N. S., Fedulova, S. A., and Kostyuk, P. G. (1986). Changes in ionic mechanisms of electrical excitability of the somatic membrane of rat's dorsal root ganglion neurons during ontogenesis. Relationship between calcium channel functioning and intracellular metabolism. *Neurophysiology (Kiev)* **17**, 827–32.

Vilven, J. and Coronado, R. (1988). Opening of dihydropyridine calcium channels in skeletal muscle membranes by inositol trisphosphate. *Nature* **336**, 587–9.

Vogel, S. and Sperelakis, N. (1981). Induction of slow action potentials by microiontophoresis of cyclic AMP into heart cells. *Journal of Molecular and Cellular Cardiology* **13**, 51–64.

Walden, J., Dietrich, L. H., and Speckmann, E.-J. (1987). A long-lasting inward current induced by intracellular injection of calcium ions in identified neurons of *Helix pomatia*. *Neuroscience Letters* **76**, 53–7.

Wanke, E., *et al.* (1987). Activation of a muscarinic receptor selectively inhibits a rapidly inactivated Ca^{2+} current in rat sympathetic neurons. *Proceedings of the National Academy of Sciences USA* **84**, 4313–17.

Wei, A. and Salkoff, L. (1986). Occult *Drosophila* calcium channels and twinning of calcium and voltage-activated potassium channels. *Science* **233**, 780–2.

Weidman, S. (1955). Effects of calcium ions and local anaesthetics on electrical properties of Purkinje fibres. *Journal of Physiology (London)* **129**, 568–82.

Wen, Y., Famulski, K. S., and Carafoli, E. (1984). Ca^{2+}-dependent K^+ permeability of heart sarcolemmal vesicles: Modulation by cAMP-dependent protein kinase activity and calmodulin. *Biochemical and Biophysical Research Communications* **122**, 237–43.

Werz, M. A. and Macdonald, R. L. (1982). Opioid peptides decrease calcium-dependent action potential duration of mouse dorsal root ganglion neurons in cell culture. *Brain Research* **239**, 315–21.

Werz, M. A. and Macdonald, R. L. (1984). Dynorphin reduces calcium-dependent action potential duration by decreasing voltage-dependent calcium conductance. *Neuroscience Letters* **46**, 185–90.

Werz, M. A. and Macdonald, R. L. (1985). Barbiturates decrease voltage-dependent calcium conductance of mouse neurons in dissociated cell culture. *Molecular Pharmacology* **28**, 269–77.

Westbrook, G. L. and Mayer, M. L. (1988). Divalent cations as modulators of NMDA-receptor channels of mouse central neurons. In *Calcium and ion channel modulation*, Proceedings of a symposium honouring R. Eckert, 26 February–1 March 1987, Los Angeles, California (ed. A. D. Grinnell, D. Armstrong, and M. B. Jackson), pp. 383–93. Plenum Press, New York.

Williams, J. A. (1980). Regulation of pancreatic acinar cell function by intracellular calcium. *American Journal of Physiology* **238**, G269–G279.

Williams, J. and Davies, J. A. (1983). The involvement of 5-hydroxytryptamine in the release of dendritic dopamine from slices of rat substantia nigra. *Journal of Pharmacy and Pharmacology* **35**, 734–7.

Williams, J. T. and North, R. A. (1985). Catecholamine inhibition of calcium action potentials in rat locus coeruleus neurones. *Neuroscience* **14**, 103–9.

Wilson, D. L., Morimoto, K., Tsuda, Y., and Brown, A. M. (1983). Interaction between calcium ions and surface charge as it relates to calcium currents. *Journal of Membrane Biology* **72**, 117–30.

Wollheim, C. B. and Pozzan, T. (1984). Correlation between cytosolic free Ca^{++} and insulin release in an insulin-secreting cell line. *Journal of Biological Chemistry* **259**, 2262–7.

Woolum, J. C. and Gorman, A. L. F. (1981). Time dependence of the calcium-activated potassium current. *Biophysical Journal* **36**, 297–302.

Yaari, Y., Hamon, B., and Lux, H. D. (1987). Development of two types of calcium channels in cultured mammalian hippocampal neurons. *Science* **235**, 680–2.

Yamaguchi, D. T., Kleeman, C. R., and Muallem, S. (1987). Protein kinase C-activated calcium channel in the osteoblast-like clonal osteosarcoma cell line UMR-106. *Journal of Biological Chemistry* **262**, 14967–73.

Yamashita, N. (1982). Enhancement of ionic currents through voltage-gated channels in the mouse oocyte after fertilization. *Journal of Physiology (London)* **329**, 263–80.

Yatani, A., Wilson, D. L., and Brown, A. M. (1983). Recovery of Ca-currents from inactivation: The roles of Ca-influx, membrane potential, and cellular metabolism. *Cellular and Molecular Neurobiology* **3**, 381–95.

Yellen, G. (1982). Single Ca^{2+}-activated nonselective cation channels in neuroblastoma. *Nature* **296**, 357–9.

Yoshida, S. (1982). Na and Ca spikes produced by ions passing through Ca channels in mouse ovarian oocytes. *Pfluegers Archiv* **395**, 84–6.

Yoshida, S. and Matsuda, Y. (1980). Responses dependent on alkaline earth cations (Ca, Sr, Ba) in dorsal root ganglion cells of the adult mouse. *Brain Research* **188**, 593–7.

Yoshida, S., Matsuda, Y., and Samejima, A. (1978). Tetrodotoxin-resistant sodium and calcium components of action potentials in dorsal root ganglion cells of the adult mouse. *Journal of Neurophysiology* **41**, 1096–106.

Yoshii, M., Tsunoo, A., and Narahashi, T. (1988). Gating and permeation properties of two types of calcium channels in neuroblastoma cells. *Biophysical Journal* **54**, 885–95.

Yue, D. and Marban, E. (1990). Permeation in the dihydropyridine-sensitive calcium channel. Multi-ion occupancy but no anomalous mole-fraction effect between Ba^{++} and Ca^{++}. *Journal of General Physiology* **95**, 911–39.

Zucker, R. S. (1989). Short-term synaptic plasticity. *Annual Review of Neuroscience* **12**, 13–31.

Zucker, R. S. and Haydon, P. G. (1988). Membrane potential has no direct role in evoking neurotransmitter release. *Nature* **335**, 360–2.

Index